FOUNDATION OF PROBABILITY THEORY

FOUNDATION OF PROBABILITY THEORY

Feng-Yu Wang
Tianjin University, China

Yong-Hua Mao
Beijing Normal University, China

Published by

World Scientific Publishing Co. Pte. Ltd.
5 Toh Tuck Link, Singapore 596224
USA office: 27 Warren Street, Suite 401-402, Hackensack, NJ 07601
UK office: 57 Shelton Street, Covent Garden, London WC2H 9HE

Library of Congress Control Number: 2024044604

British Library Cataloguing-in-Publication Data
A catalogue record for this book is available from the British Library.

概率论基础
Originally published in Chinese by Beijing Normal University Press (Group) Co., Ltd.

FOUNDATION OF PROBABILITY THEORY

ISBN 978-981-12-9885-1 (hardcover)
ISBN 978-981-98-0035-3 (ebook for institutions)
ISBN 978-981-98-0036-0 (ebook for individuals)

For any available supplementary material, please visit
https://www.worldscientific.com/worldscibooks/10.1142/14003#t=suppl

Desk Editors: Nambirajan Karuppiah/Rok Ting Tan

Typeset by Stallion Press
Email: enquiries@stallionpress.com

Preface

Why shall we learn the course "Foundation of Probability Theory" after the elementary course "Probability Theory"? The reason is that the elementary probability theory describes specific distributions induced by random trials, which is intuitively clear but mathematically less rigorous, while the foundation of probability theory established by Kolmogorov is an axiomatization theory, which makes probability theory a rigorous branch of mathematics.

For example, in the elementary probability theory, the sample space is the total of the possible results appearing in a random trial, and each subset of this space is called an event, whose probability is defined as the limit of its appearing frequency as the number of the trials goes to infinity. These concepts are intuitively clear but not mathematically rigorous: Why can the trial be repeated infinite times? Why must the frequency converge? And how can one fix the limit if it does converge? One may argue that this limit exists due to the law of large numbers. However, the law of large numbers itself is established based on the definition of probability, which leads to a circular argument.

Now, the motivation to learn the course becomes clear: it enables us to grasp a serious foundation of probability theory in the mathematical axiomatic system. Contrary to the elementary probability theory which deals with random events in specific examples of random trials, *Foundation of Probability Theory* is a general mathematical theory which provides rigorous descriptions of these examples. Therefore, this course has all characters of mathematical theories: abstract contents, extensive applications, complete structures, and clear conclusions. Due to abstract contents, we will face many difficulties during learning. To overcome these difficulties, a crucial trick is to keep in mind those concrete examples while trying to understand an abstract context and compare the abstract theory with related courses learned before, especially with the Lebesgue measure theory. In the following, we give a brief chapterwise summary of the main contents of this textbook.

To define events without random trials, we first fix a global set Ω and then construct a class $\mathscr{A}$ of subsets of Ω, which is equipped with an algebra structure, so that each element in $\mathscr{A}$ is measurable in a reasonable way. We then call the couple $(\Omega, \mathscr{A})$ a measurable space, where Ω refers to the sample space of a random trial and $\mathscr{A}$ stands for the set of events. In general, $\mathscr{A}$ is strictly smaller than the class of all subsets of Ω, i.e. not all subsets of Ω are measurable. For instance, there exist nonmeasurable sets in the Lebesgue measure theory. Following the line of the Lebesgue measure theory, we assume that $\mathscr{A}$ contains Ω and is closed under countable set operations, which leads to the concept of σ-algebra. Furthermore, the probability of an event can be considered as the nonnegative survey results for sets in $\mathscr{A}$, which is this a function $\mathbb{P} : \mathscr{A} \to [0, \infty)$. According to the requirement of probability measure in the probability theory, we postulate that $\mathbb{P}(\Omega) = 1$ and $\mathbb{P}$ has σ-additivity, i.e. $\mathbb{P}\left(\bigcup_{n=1}^{\infty} A_n\right) = \sum_{n=1}^{\infty} \mathbb{P}(A_n)$ for a sequence of mutually disjoint sets $\{A_n\}_{n\geqslant 1} \subset \mathscr{A}$. Without the restriction $\mathbb{P}(\Omega) = 1$, the map $\mathbb{P}$ is called a measure and is denoted by μ rather than $\mathbb{P}$ to emphasize the difference. In this way, we construct a triple $(\Omega, \mathscr{A}, \mathbb{P})$, or more generally $(\Omega, \mathscr{A}, \mu)$, which is called a probability space or a measure space.

So, how can we construct a probability measure on a σ-algebra? According to the Lebesgue measure theory, we first define the measure value for simple sets, for instance, the intervals, and then extend it to all measurable sets by an extension argument. To abstract this method in the general framework, we first introduce the semi-algebra for subsets of Ω in terms of the property of right semi-closed intervals in the Euclidean space. From this semi-algebra of sets, we generate the σ-algebra $\mathscr{A}$ by establishing the monotone class theorem. Moreover, by the monotone class theorem and the construction ideas of Lebesgue measure, a measure defined on a semi-algebra $\mathscr{S}$ can be (under a σ-finite condition uniquely) extended to the minimal σ-algebra generated by $\mathscr{S}$. This is known as the *measure extension theorem,* the core result of Chapter 1.

Having the measure space $(\Omega, \mathscr{A}, \mu)$ in hand, the tasks in Chapters 2 and 3 are to survey a *measurable function*, and the value is called the integral of the function with respect to μ. The definition and properties of integrals are inhered from the theory of Lebesgue integrals and hence are easy to understand with a basis of the Lebesgue measure theory. In particular, on a probability space $(\Omega, \mathscr{A}, \mathbb{P})$, a measurable function is called a random variable, whose expectation is defined as the integral with respect to the probability measure. By the integral transformation formula (Lebesgue–Stieltjes integral expression), the expectation can be formulated as integral of the identity function with respect to the distribution of the random variable, where the distribution is a probability measure on the real line. In order to classify the distributions of random variables, we consider the decomposition of measures in Chapter 3.

To study several or infinite many random variables together, we introduce product probability spaces and consider the conditional properties of some random variables given other ones. These are treated in Chapters 4 and 5, where the main difficulty is to clarify the definition of conditional expectation given a sub σ-algebra and to introduce the regular conditional probability which enables one to construct measures on product spaces which is fundamental for further study of stochastic processes.

Chapter 6 presents several equivalent definitions of the weak convergence for finite measures, which are also equivalent to the convergence of the characteristic functions for finite measures in the multi-dimensional Euclidean space. Chapter 7 introduces some probability distances on the space of probability measures. Both chapters are important to develop the limit theory of random variables and stochastic processes.

Finally, Chapter 8 introduces derivatives for functions of finite measures, establishes the chain rule and derivative formulas. These provide a quick way for readers to enter the frontier of analysis and geometry on the space of measures.

In conclusion, this course is an abstract rigorization of the elementary probability theory. Key points include the monotone class theorem, measure extension theorem, conditional expectation and regular conditional probability, and the weak convergence. To make the whole book easy to follow, in the beginning of each part, we briefly introduce the main purpose of study based on previous contents, figure out the main structure, and explain the key idea of study. If one understands clearly the backgrounds and basic ideas for the study of each part, it is not hard to grasp the whole contents of this textbook. There are many books containing these contents, see an incomplete list of references in the end of this book.

The first seven chapters of the textbook are translated and modified from the Chinese version published in 2010 by the Beijing Normal University Press. We would like to thank the Executive Editor Ms Fengjuan Liu for the encouragement and the efficient work. We gratefully acknowledge the support from the National Key R&D Program of China (2022YFA1006000, 2020YFA0712900) and the National Nature Science Foundation of China (11921001).

About the Authors

Feng-Yu Wang is a Professor at Tianjin University. After receiving his Ph.D. in 1993 from Beijing Normal University, he was exceptionally appointed by Beijing Normal University in 1995 as a Full Professor by the Educational Ministry of China in 2000 as a reputed Chang-Jiang Chair, by Swansea University in 2007 as a Research Chair, and by Tianjin University in 2016 as a Chair Professor. His research areas include stochastic analysis on manifolds, functional inequalities and applications, stochastic (partial) differential equations, distribution dependent stochastic differential equations, and Wasserstein convergence of empirical measures. He served as an Associated Editor of *Electronic Journal of Probability, Electronic Communications in Probability, Journal of Theoretical Probability, Communications in Pure and Applied Analysis, Science in China Mathematics*, and *Frontiers of Mathematics in China.*

Yong-Hua Mao is a Professor at Beijing Normal University. He has worked on Markov chains, jump processes, and other Markov processes, especially on stationary and quasi-stationary for Markov processes. He served as an Associate Editor of *Statistics and Probability Letters* and the *Chinese Journal of Applied Probability and Statistics.*

Contents

Chapter 1

Class of Sets and Measure

What is a measure? It is a tool to determine the weights of "measurable sets" which satisfy the countable additivity property, i.e. the sum of weights for countable many disjoint sets coincides with the weight of the union of these sets. For example, under the Lebesgue measure, the weight of an interval $[a, b)$ for real numbers $b \geqslant a$ is its length $b-a$, which uniquely determines a measure on the class of "Lebesgue measurable sets" on $\mathbb{R}$. The aim of this chapter is to choose a reasonable class of subsets (i.e. measurable sets) for a given global set Ω and to construct a measure on this class. To this end, we first generate a class of subsets sharing the following features of Lebesgue measurable sets: (1) it contains the empty set and the total set; (2) it is closed under the countably infinite set operations (set union, intersection and difference). A class of sets with these properties is called a σ-algebra or σ-field, which is our ideal class of "measurable subsets" of the abstract global set Ω. To define a measure on the σ-algebra, let us again go back to the Lebesgue measure on $\mathbb{R}$.

As mentioned above, the Lebesgue measure of an interval is defined as the length. By a natural extension procedure, this measure can be extended to the smallest σ-algebra containing intervals, which is nothing but the Borel σ-algebra whose completion is a class of Lebesgue measurable sets. To realize the same procedure for the present abstract setting with Ω in place of $\mathbb{R}$, we consider the "semi-algebra" which is a class of sets sharing the following features of intervals: (1) it contains the empty set and the total set; (2) it is closed under the set intersection; (3) the difference of any two sets can be expressed by the union of finite disjoint sets in the class. We first induce the smallest σ-algebra from the semi-algebra, where the main tool is called the monotone class theorem, and then extend a measure from the semi-algebra to the induced σ-algebra, where the key step is to establish the measure extension theorem. These two theorems are the key results of this chapter.

1.1 Class of Sets and Monotone Class Theorem

1.1.1 Semi-algebra

We first introduce operations for subsets of the global set Ω. Let $\cup$ and $\cap$ denote the union and intersection, respectively, let A^c be the complement of the set A, and let $A - B := A \cap B^c$ be the difference of A and B, which is called a proper difference if $B \subset A$. For simplicity, we will use AB to stand for $A \cup B$, $A + B$ for $A \cup B$ with $AB = \varnothing$, and $\sum_n A_n$ for the union of finite or countable many disjoint sets $\{A_n\}$.

Then the semi-algebra of sets is defined in terms of the above-mentioned features of intervals.

Definition 1.1. A class $\mathscr{S}$ of subsets of Ω is called a semi-algebra (of sets) in Ω if

(1) $\Omega, \varnothing \in \mathscr{S}$,
(2) $A \cap B \in \mathscr{S}$ for $A, B \in \mathscr{S}$,
(3) for $A_1, A \in \mathscr{S}$ with $A_1 \subset A$, there exist $n \geqslant 1$ and $A_1, A_2, \ldots, A_n \in \mathscr{S}$ disjoint mutually such that $A = \sum_{i=1}^{n} A_i$.

Property 1.2. Under items (1) and (2) in Definition 1.1, item (3) is equivalent to the following:

(3′) if $A \in \mathscr{S}$, then $\exists n \geqslant 1$ and $A_1, A_2, \ldots, A_n \in \mathscr{S}$ mutually disjoint such that $A^c = \sum_{i=1}^{n} A_i$.

Proof. (3) $\Rightarrow$ (3′): Since $A \subset \Omega$, it follows from (3) that $\exists n \geqslant 1$ and $A_1, A_2, \ldots, A_n \in \mathscr{S}$ disjoint mutually, which are all disjoint with A, such that $\Omega = A + \sum_{i=1}^{n} A_i$, so $A^c = \sum_{i=1}^{n} A_i$.

$(3') \Rightarrow (3)$: It follows from $(3')$ that $\exists n \geqslant 2$ and $A_2, \ldots, A_n \in \mathscr{S}$ mutually disjoint such that $A_1^c = \sum_{i=2}^{n} A_i$, so $A = A_1 + \sum_{i=2}^{n} A_i \cap A$. □

Example 1.3. Let $\Omega = [0, +\infty)$, $\mathscr{S} = \{[a, b) : 0 \leqslant a \leqslant b \leqslant +\infty\}$. Then $\mathscr{S}$ is a semi-algebra in Ω.

To induce the σ-algebra from a semi-algebra, we introduce a relay notion "algebra", which is closed under finite many operations.

1.1.2 Algebra

Definition 1.4. A class $\mathscr{F}$ of subsets of Ω is called an algebra (of sets) in Ω, or Boolean algebra in Ω, if

(1) $\Omega \in \mathscr{F}$,
(2) $A, B \in \mathscr{F}$ implies $A - B \in \mathscr{F}$.

Property 1.5. Under item (1) in Definition 1.4, item (2) is equivalent to any one of the following:

(2′) $A, B \in \mathscr{F}$ implies $A \cup B, A^c, B^c \in \mathscr{F}$,
(2″) $A, B \in \mathscr{F}$ implies $A \cap B, A^c, B^c \in \mathscr{F}$.

Proof. We will prove $(2'') \Rightarrow (2') \Rightarrow (2) \Rightarrow (2'')$.

$(2'') \Rightarrow (2')$: It follows from $(2'')$ that $\mathscr{F}$ is closed under complement and intersection, so that $A, B \in \mathscr{F}$ implies $A \cup B = (A^c \cap B^c)^c \in \mathscr{F}$.
$(2') \Rightarrow (2)$: Assume $A, B \in \mathscr{F}$. It follows from $(2')$ that $\mathscr{F}$ is closed under complement and union, so that $A - B = (A^c \cup B)^c \in \mathscr{F}$.
$(2) \Rightarrow (2'')$: Assume $A, B \in \mathscr{F}$. It follows from (2) that $A^c = \Omega - A$, $B^c = \Omega - B \in \mathscr{F}$, so that $A \cap B = A - B^c \in \mathscr{F}$.

□

Proposition 1.6. *If $\mathscr{F}$ is an algebra in Ω, then $\forall A, B \in \mathscr{F}$ and we have $A^c, B^c, A \cap B, A \cup B, A - B \in \mathscr{F}$.*

Obviously, an algebra is a semi-algebra. The following theorem provides an explicit formulation of the induced algebra from a semi-algebra.

Theorem 1.7. *If $\mathscr{S}$ is a semi-algebra, then*

$$\mathscr{F} := \left\{ \sum_{k=1}^{n} A_k : n \geqslant 1, A_k \in \mathscr{S}(1 \leqslant k \leqslant n) \text{ are mutually disjoint} \right\}$$

is the smallest algebra containing $\mathscr{S}$, which is called the algebra induced (or generated) by $\mathscr{S}$, and is denoted by $\mathscr{F}(\mathscr{S})$.

Proof. First, we prove that $\mathscr{F}$ is an algebra. Obviously, item (1) in Definition 1.4 is fulfilled. Moreover, $\forall A, B \in \mathscr{F}, \exists A_1, A_2, \ldots, A_n \in \mathscr{S}$ and $B_1, B_2, \ldots, B_m \in \mathscr{S}$, mutually disjoint, respectively, such that $A = \sum_{i=1}^{n} A_i$, $B = \sum_{i=1}^{m} B_i$. Then $A \cap B = \sum_{i,j} A_i \cap B_j$. It follows from Definition 1.1(2) that $A \cap B \in \mathscr{F}$, so that $\mathscr{F}$ is closed under finite intersections.

Next, by Property 1.5, to prove that $\mathscr{F}$ is an algebra in Ω, we only need to verify $A^c \in \mathscr{F}$ for any $A \in \mathscr{F}$. Let $A = \sum_{i=1}^{n} A_i \in \mathscr{F}, A_i \in \mathscr{S}$. Then $A^c = \bigcap_{i=1}^{n} A_i^c$. By Property 1.2, we see that A_i^c can be expressed by the union of mutually disjoint sets in $\mathscr{S}$, so $A_i^c \in \mathscr{F}$. Since $\mathscr{F}$ is closed under finite many intersections, we obtain $A^c \in \mathscr{F}$.

Finally, for any algebra $\mathscr{F}' \supset \mathscr{S}$, Property 1.5 implies $\mathscr{F}' \supset \mathscr{F}$. □

Example 1.8. $\mathscr{S}$ in Example 1.3 is not an algebra, and by Theorem 1.7, its induced algebra is

$$\mathscr{F}(\mathscr{S}) = \left\{ \sum_{i=1}^{n} [a_i, b_i) : n \geqslant 1, 0 \leqslant a_1 \leqslant b_1 \leqslant a_2 \leqslant b_2 \leqslant \cdots \leqslant a_n \leqslant b_n \right\}.$$

1.1.3 σ-algebra

According to the property of Lebesgue measurable sets, a σ-algebra should be closed under the countable many operations of sets. Since the union and intersection of sets are dual to each other by complement, it suffices to have the closedness by complement and countable many unions.

Definition 1.9. A class $\mathscr{A}$ of subsets of Ω is called a σ-algebra (or σ-field) in Ω if

(1) $\Omega \in \mathscr{A}$,
(2) $A^c \in \mathscr{A}$ holds for $A \in \mathscr{A}$,
(3) $\bigcup_{n=1}^{\infty} A_n \in \mathscr{A}$ holds for any $\{A_n\}_{n\geqslant 1} \subset \mathscr{A}$.

In this case, we call $(\Omega, \mathscr{A})$ a measurable space, and each element in $\mathscr{A}$ is called an $\mathscr{A}$-measurable set or simply a measurable set.

Property 1.10. A σ-algebra is an algebra.

Property 1.11. Under items (1) and (2) in Definition 1.9, (3) is equivalent to

(3′) $\bigcap_{n=1}^{\infty} A_n \in \mathscr{A}$ for $A_n \in \mathscr{A}, n = 1, 2, \ldots$

Proof. Note that $\bigcap_{n=1}^{\infty} A_n = \left(\bigcup_{n=1}^{\infty} A_n^c\right)^c$. □

Property 1.12. The intersection of a family of σ-algebras in Ω is also a σ-algebra.

Proof. Let $\{\mathscr{A}_r : r \in \Gamma\}$ be a family of σ-algebras in Ω. Then $\mathscr{A} = \bigcap_{r\in\Gamma} \mathscr{A}_r$ is a σ-algebra in Ω as well because of the following:

(1) for any $r \in \Gamma$, we have $\varnothing, \Omega \in \mathscr{A}_r$, so that $\varnothing, \Omega \in \mathscr{A}$;
(2) if $A \in \mathscr{A}$, then $A \in \mathscr{A}_r$ for any $r \in \Gamma$, so that $A^c \in \mathscr{A}_r$ $(r \in \Gamma)$, i.e. $A^c \in \mathscr{A}$;
(3) if $A_1, A_2, \ldots \in \mathscr{A}$, then $A_1, A_2, \ldots \in \mathscr{A}_r$ for any $r \in \Gamma$, so that $\bigcup_{n=1}^{\infty} A_n \in \mathscr{A}_r$ for all $r \in \Gamma$, hence $\bigcup_{n=1}^{\infty} A_n \in \mathscr{A}$. □

Example 1.13. $\mathscr{A} = \{\varnothing, \Omega\}$ is the smallest σ-algebra in Ω, while

$$\mathscr{A} = 2^{\Omega} := \{A : A \subset \Omega\}$$

is the largest σ-algebra in Ω, where the notation 2^{Ω} comes from the fact that a subset A of Ω is identified with the element in $\{0,1\}^{\Omega}$: $\Omega \ni \omega \mapsto \mathbf{1}_A(\omega)$, where $\mathbf{1}_A$ is the indicator function of A.

Theorem 1.14. *Let $\mathscr{C}$ be a class of subsets of Ω. Then there exist a unique σ-algebra $\mathscr{A}$ in Ω such that*

(1) $\mathscr{C} \subset \mathscr{A}$,
(2) *if $\mathscr{A}$ is a σ of Ω and $\mathscr{A} \supset \mathscr{C}$, then $\mathscr{A} \supset \mathscr{A}$.*

We denote $\mathscr{A}$ by $\sigma(\mathscr{C})$ and call it the σ-algebra induced (or generated) from $\mathscr{C}$.

Proof. Since the largest σ-algebra includes $\mathscr{C}$, there exists at least one σ-algebra including $\mathscr{C}$. Let $\mathscr{A}$ be the intersection of all σ-algebras including $\mathscr{C}$. By Property 1.12, $\mathscr{A}$ is the smallest σ-algebra including $\mathscr{C}$. □

The following theorem shows that the induced procedure from a semi-algebra to σ-algebra can be decomposed into two steps, i.e. first induce the algebra and then the σ-algebra.

Theorem 1.15. *If $\mathscr{S}$ is a semi-algebra of Ω, then $\sigma(\mathscr{S}) = \sigma(\mathscr{F}(\mathscr{S}))$.*

Proof. Since $\sigma(\mathscr{F}(\mathscr{S})) \supset \mathscr{S}$, we have $\sigma(\mathscr{F}(\mathscr{S})) \supset \sigma(\mathscr{S})$. Conversely, since $\sigma(\mathscr{S})$ is an algebra including $\mathscr{S}$, we have $\sigma(\mathscr{S}) \supset \mathscr{F}(\mathscr{S})$, and hence, $\sigma(\mathscr{S}) \supset \sigma(\mathscr{F}(\mathscr{S}))$. □

Example 1.16. Let $(\Omega, \mathscr{T})$ be a topology space where $\mathscr{T}$ is the class of all open subsets of Ω. The σ-algebra $\mathscr{B} := \sigma(\mathscr{T})$ is called Borel field (or Borel σ-algebra).

1.1.4 Monotone class theorem

By Theorem 1.7, the algebra is easily induced from a semi-algebra. Combining this with Theorem 1.15, to induce the σ-algebra from a semi-algebra, one only needs to generate it from the induced algebra. Note that the difference between the algebra and the σ-algebra is the following: the former is only closed under finite many operations, while the latter is closed under countably infinite many operations. Intuitively, countably infinite many operations can be characterized as the limit of finite many operations. So, it is reasonable to consider the limit for sequences of sets.

Note that the limit for a sequence of sets is defined only in the monotone case by the union (respectively, intersection) for an increasing (respectively, decreasing) sequence. This leads to the notion of monotone class.

Definition 1.17. A class $\mathscr{M}$ of subsets of Ω is called a monotone class if it is closed for the limits of monotone sequences, that is,

(1) if $A_n \in \mathscr{M}, n = 1, 2, \ldots$, and $A_1 \subset A_2 \subset \ldots$, then $\bigcup_{n=1}^{\infty} A_n \in \mathscr{M}$;

(2) if $A_n \in \mathscr{M}, n = 1, 2, \ldots$, and $A_1 \supset A_2 \supset \ldots$, then $\bigcap_{n=1}^{\infty} A_n \in \mathscr{M}$.

Theorem 1.18. *A class of subsets of Ω is a σ-algebra if and only if it is both an algebra and a monotone class.*

Proof. It suffices to prove the sufficiency. Let $\mathscr{A}$ be both an algebra and a monotone class. We only need to show that $\mathscr{A}$ is closed under countably many unions. Since $\mathscr{A}$ is an algebra, $\forall A_1, A_2, \ldots \in \mathscr{F}$, $B_n := \bigcup_{i=1}^{n} A_i \in \mathscr{A}$ is increasing in n. Since $\mathscr{A}$ is a monotone class, this implies

$$\bigcup_{i=1}^{\infty} A_i = \bigcup_{n=1}^{\infty} B_n \in \mathscr{A}.$$

Then the proof is finished. □

Theorem 1.19. *Let $\mathscr{C}$ be an algebra in Ω. Then there exists a unique monotone class $\mathscr{M}_0$ in Ω fulfilling*

(1) $\mathscr{M}_0 \supset \mathscr{C}$,
(2) $\mathscr{M} \supset \mathscr{M}_0$ *holds for any monotone class* $\mathscr{M}$ *including* $\mathscr{C}$.

We call $\mathscr{M}_0$ the monotone class induced (or generated) from $\mathscr{C}$ and denote it by $\mathscr{M}(\mathscr{C})$.

Theorem 1.20. *Let $\mathscr{F}$ be an algebra. Then $\mathscr{M}(\mathscr{F}) = \sigma(\mathscr{F})$.*

Proof. By Theorem 1.18, it suffices to prove that $\mathscr{M}(\mathscr{F})$ is an algebra.

(a) We first show that $A \in \mathscr{M}(\mathscr{F}) \Rightarrow A^c \in \mathscr{M}(\mathscr{F})$. Let $\mathscr{M}_1 = \{A : A^c \in \mathscr{M}(\mathscr{F})\}$. For any decreasing sequence $\{A_n\} \subset \mathscr{M}_1$, we have $\{A_n^c\} \subset \mathscr{M}(\mathscr{F})$, which is increasing. Since $\mathscr{M}(\mathscr{F})$ is a monotone class, we have $\bigcup_{n=1}^{\infty} A_n^c \in \mathscr{M}(\mathscr{F})$, so $\bigcap_{n=1}^{\infty} A_n = \left(\bigcup_{n=1}^{\infty} A_n^c\right)^c \in \mathscr{M}_1$. Similarly, we see that $\mathscr{M}_1$ is closed under the limits of increasing sequences. Thus, $\mathscr{M}_1$ is a monotone class including $\mathscr{F}$, so it includes $\mathscr{M}(\mathscr{F})$. Thus, $A^c \in \mathscr{M}(\mathscr{F})$ for any $A \in \mathscr{M}(\mathscr{F})$.

(b) Next, we prove that for any $A \in \mathscr{F}$, $A \bigcap B \in \mathscr{M}(\mathscr{F})$ holds for $B \in \mathscr{M}(\mathscr{F})$. To this end, let $\mathscr{M}_A = \{B \in \mathscr{M}(\mathscr{F}) : A \cap B \in \mathscr{M}(\mathscr{F})\}$. Then $\mathscr{M}_A \supset \mathscr{F}$. If $\{B_n\} \subset \mathscr{M}_A$ is increasing, then so is $\{A \bigcap B_n\} \subset \mathscr{M}(\mathscr{F})$. Since $\mathscr{M}(\mathscr{F})$ is a monotone class, we obtain $A \cap \left(\bigcup_{n=1}^{\infty} B_n\right) \in \mathscr{M}(\mathscr{F})$, which implies $\bigcup_{n=1}^{\infty} B_n \in \mathscr{M}_A(\mathscr{F})$. Similar argument shows that $\mathscr{M}_A$ is a monotone class. Thus, $\mathscr{M}_A \supset \mathscr{M}(\mathscr{F})$; i.e. $A \cap B \in \mathscr{M}(\mathscr{F})$ holds for any $B \in \mathscr{M}(\mathscr{F})$.

(c) Finally, we prove $A \cap B \in \mathscr{M}(\mathscr{F})$ for $A, B \in \mathscr{M}(\mathscr{F})$. By (b), $\mathscr{M}_A$ is a monotone class including $\mathscr{F}$, so that $\mathscr{M}_A \supset \mathscr{M}(\mathscr{F})$. Thus, $A \cap B \in \mathscr{M}(\mathscr{F})$. □

The trick behind the proof can be summarized as follows. To prove that a class $\mathscr{C}_1$ of sets has certain property, we define a new class $\mathscr{C}_2$ consisting of all sets having this property and then it suffices to show that $\mathscr{C}_1 \subset \mathscr{C}_2$. To realize this procedure, we sometimes need to split it into several steps, as we have done above with two steps. This technique will be used frequently.

The monotonicity is easier to check than the closedness under countable unions (or intersections). In the spirit of Theorem 1.18 that the monotone class and algebra give rise to σ-algebra, in the following, we introduce another pair of classes to form σ-algebra.

Definition 1.21. (1) A class $\mathscr{C}$ of subsets of Ω is called a π-system if it is closed under intersections.

(2) A class $\mathscr{C}$ of subsets of Ω is called a λ-system if it fulfills the following:

(i) $\Omega \in \mathscr{C}$,
(ii) $B - A \in \mathscr{C}$ holds for $A, B \in \mathscr{C}, A \subset B$,
(iii) $\bigcup_{n=1}^{\infty} A_n \in \mathscr{C}$ holds for any increasing $\{A_n\} \subset \mathscr{C}$.

Property 1.22. If $\mathscr{C}$ is a λ-system, then it is a monotone class.

Proof. If $\{A_n\} \subset \mathscr{C}$ is decreasing, then $\{A_n^c\} \subset \mathscr{C}$ is increasing. By the definition of λ-system, we obtain $\bigcup_{n=1}^{\infty} A_n^c \in \mathscr{C}$, so that $\bigcap_{n=1}^{\infty} A_n = \Omega - \bigcup_{n=1}^{\infty} A_n^c \in \mathscr{C}$. □

Property 1.23. If $\mathscr{C}$ is both a π-system and a λ-system, then $\mathscr{C}$ is a σ-algebra.

Proof. By Theorem 1.18 and Property 1.22, it suffices to prove that $\mathscr{C}$ is an algebra, which follows easily from Definition 1.21 of π-system and (ii) Definition 1.21 of λ-system. □

In the same spirit of Theorems 1.14 and 1.19 for the induced σ-algebra and monotone class, any class of sets $\mathscr{C}$ induces a unique λ-system $\lambda(\mathscr{C})$, which is the smallest λ-system including $\mathscr{C}$.

Theorem 1.24. *Let $\mathscr{C}$ be a π-system. Then $\lambda(\mathscr{C}) = \sigma(\mathscr{C})$.*

Proof. Since $\lambda(\mathscr{C}) \subset \sigma(\mathscr{C})$ and $\lambda(\mathscr{C})$ is a monotone class, by Theorem 1.18, it suffices to prove $\lambda(\mathscr{C})$ is an algebra. By the definition of λ-system, $\lambda(\mathscr{C})$ is closed under complement, so it remains to verify the closedness under intersections. We split the proof into two steps:

(1) Let $A \in \lambda(\mathscr{C}), B \in \mathscr{C}$. We intend to prove that $A \cap B \in \lambda(\mathscr{C})$. By the trick explained above, let $\mathscr{C}_B = \{A : A \cap B \in \lambda(\mathscr{C})\}$. Since $\mathscr{C}$ is a π-system, we have $\mathscr{C}_B \supset \mathscr{C}$. By the definition of $\mathscr{C}_B$ and the fact $\lambda(\mathscr{C})$ is a λ-system, it is easy to verify that $\mathscr{C}_B$ is also a λ-system, so $\mathscr{C}_B \supset \lambda(\mathscr{C})$, that is, $A \cap B \in \lambda(\mathscr{C})$ for any $A \in \lambda(\mathscr{C})$.
(2) Let $B \in \lambda(\mathscr{C})$. From (1), we see that $\mathscr{C}_B \supset \mathscr{C}$ and it is a λ-system, so that $\mathscr{C}_B \supset \lambda(\mathscr{C})$. Thus, $A \cap B \in \lambda(\mathscr{C})$ for $A, B \in \lambda(\mathscr{C})$. □

Having the above preparations, we obtain the following important theorem.

Theorem 1.25 (Monotone class theorem). *Let $\mathscr{C}$ and $\mathscr{A}$ be two classes of subsets of Ω with $\mathscr{C} \subset \mathscr{A}$.*

(1) *If $\mathscr{A}$ is a λ-system and $\mathscr{C}$ is a π-system, then $\sigma(\mathscr{C}) \subset \mathscr{A}$.*
(2) *If $\mathscr{A}$ is a monotone class and $\mathscr{C}$ is an algebra, then $\sigma(\mathscr{C}) \subset \mathscr{A}$.*

This is the main result for classes of sets. In the following, we explain the main idea to apply the monotone class theorem. Let $\mathscr{C}$ be a class of sets having certain property S, one wants to verify the same property for sets in $\sigma(\mathscr{C})$. For this, let $\mathscr{A} = \{B : B \text{ has property } S\}$, so that $\mathscr{A} \supset \mathscr{C}$. By the monotone class theorem, it suffices to show that $\mathscr{C}$ is a π-system or an algebra, and accordingly, $\mathscr{A}$ is a λ-system or a monotone class.

Remark 1.26. The following diagram summarizes the relations of various classes of sets:

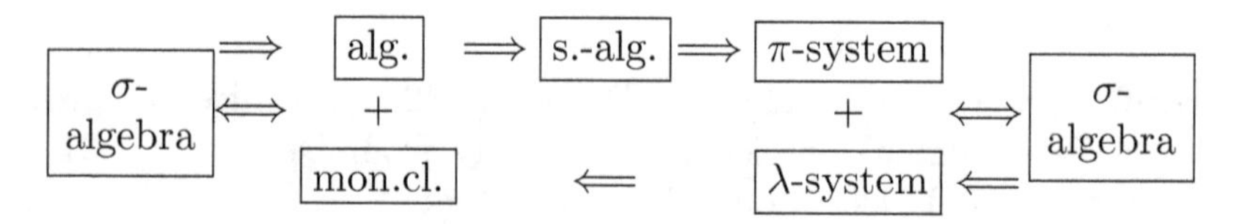

where alg. stands for algebra, mon.cl. for monotone class, and s.-alg. for semi-algebra.

1.1.5 Product measurable space

Let $(\Omega_i, \mathscr{A}_i), i = 1, \ldots, n$ be finite many measurable spaces. Let

$$\mathscr{C} = \{A_1 \times \cdots \times A_n : A_i \in \mathscr{A}_i,\ 1 \leqslant i \leqslant n\},$$

where each element in $\mathscr{C}$ is called a rectangle in the product space

$$\Omega := \Omega_1 \times \cdots \times \Omega_n.$$

It is easy to check that $\mathscr{C}$ is a semi-algebra in Ω. We call $\mathscr{A} := \sigma(\mathscr{C})$ the product σ-algebra of $\mathscr{A}_1, \ldots, \mathscr{A}_n$ and denote it by $\mathscr{A}_1 \times \cdots \times \mathscr{A}_n$. Moreover, $(\Omega, \mathscr{A})$ is called the product measurable space of $(\Omega_i, \mathscr{A}_i), i = 1, \ldots, n$.

Theorem 1.27 (Associative law). $\forall n \geqslant 3, 1 \leqslant k \leqslant n$, *we have*

$$\mathscr{A}_1 \times \mathscr{A}_2 \times \cdots \times \mathscr{A}_n = (\mathscr{A}_1 \times \cdots \times \mathscr{A}_k) \times (\mathscr{A}_{k+1} \times \cdots \times \mathscr{A}_n).$$

Theorem 1.27 can be derived by the definition of product σ-algebra and the monotone class theorem, whose proof is left as an exercise.

1.2 Measure

After constructing a measurable space $(\Omega, \mathscr{A})$, we intend to define a real function on $\mathscr{A}$, which is called a measure if it is nonnegative and satisfies the countable additivity. In general, we consider a real function defined on a class of sets, which is called a function of sets.

1.2.1 Function of sets

Definition 1.28. A function on a class of sets $\mathscr{C}$ in Ω is a map

$$\Phi : \mathscr{C} \to (-\infty, +\infty],$$

such that $\Phi(A) < \infty$ for some $A \in \mathscr{C}$.

We allow a function of sets to take value $+\infty$ such that the Lebesgue measure is included. On the other hand, we do not allow it to take value $-\infty$ to avoid the sum of $+\infty$ and $-\infty$ when the additivity property is considered. In general, we study functions of sets with the following properties:

(1) (*Additivity*) A function Φ of sets is called additive if $\Phi(A+B) = \Phi(A)+\Phi(B)$ holds for any disjoint $A, B \in \mathscr{C}$ such that $A+B \in \mathscr{C}$.
(2) (*Finite additivity*) A function Φ of sets is called finitely additive if

$$\Phi\left(\sum_{i=1}^{n} A_i\right) = \sum_{i=1}^{n} \Phi(A_i)$$

holds for any $n \geqslant 2$, and $A_1, \ldots, A_n \in \mathscr{C}$ mutually disjoint with $\sum_{i=1}^{n} A_i \in \mathscr{C}$.
(3) (*σ-additivity, or countable additivity*) A function Φ of sets is called σ-additive if

$$\Phi\left(\sum_{i=1}^{\infty} A_i\right) = \sum_{i=1}^{\infty} \Phi(A_i)$$

holds for any $\{A_n\}_{n\geqslant 1} \subset \mathscr{C}$ mutually disjoint with $\sum_{i=1}^{\infty} A_i \in \mathscr{C}$.

(4) (*Finiteness*) A function Φ of sets is called finite if $\Phi(A) \in \mathbb{R}$ holds for all $A \in \mathscr{C}$.
(5) (σ-*finiteness*) A function Φ of sets is called σ-finite if for any $A \in \mathscr{C}$, there exists a sequence $\{A_n\}_{n\geqslant 1} \subset \mathscr{C}$ such that $\Phi(A_n) \in \mathbb{R}$ $(\forall n \geqslant 1)$ and $A = \bigcup_{n=1}^{\infty} A_n$.

Definition 1.29. A signed measure is a function of sets with σ-additivity. A measure is a signed measure taking nonnegative values. A probability measure is a measure with $\Phi(\Omega) = 1$. If a function Φ of sets takes nonnegative values and is finitely additive, then it is call a finitely additive measure.

Note that a signed measure or a finitely additive measure may not be a measure. The following propositions for functions of sets are obvious, so the proofs are omitted.

Proposition 1.30. *Let Φ be a function on $\mathscr{C}$.*

(1) *Finite additivity $\Rightarrow$ additivity.*
(2) *If $\varnothing \in \mathscr{C}$, then σ-additivity $\Rightarrow$ finite additivity.*
(3) *If $\mathscr{C}$ is an algebra, then finite additivity $\Leftrightarrow$ additivity.*
(4) *If Φ is additive and $\varnothing \in \mathscr{C}$, then $\Phi(\varnothing) = 0$.*

The following result characterizes the properties of functions on different classes of sets.

Property 1.31.

(1) (*Subtractivity*) Let Φ be an additive function on an algebra $\mathscr{F}$, and let $A, B \in \mathscr{F}$ with $A \subset B$. We have $\Phi(B) = \Phi(A)+\Phi(B-A)$. If $\Phi(A) < \infty$, then $\Phi(B - A) = \Phi(B) - \Phi(A)$.
(2) (*Monotonicity*) Let μ be a finitely additive measure on a semi-algebra $\mathscr{S}$. Then $\mu(A) \leqslant \mu(B)$ holds for $A, B \in \mathscr{S}$ with $A \subset B$.
(3) (*Finiteness*) Let Φ be a finitely additive function on a semi-algebra $\mathscr{S}$. If $\Phi(B) < \infty$ and $A \subset B$, then $\Phi(A) < \infty$. In addition, if $\Phi(\Omega) < \infty$, then Φ is finite.
(4) (σ-*finiteness*) Let Φ be a finitely additive function on a semi-algebra $\mathscr{S}$. If $\Omega = \bigcup_{n=1}^{\infty} A_n$ with $A_n \in \mathscr{S}$ and $\Phi(A_n) < \infty$ $(\forall n \geqslant 1)$, then for $\forall A \in \mathscr{S}, \exists \{A'_n\} \subset \mathscr{S}$ mutually disjoint such that $A = \sum_{n=1}^{\infty} A'_n$ and $\Phi(A'_n) < \infty$ $(\forall n \geqslant 1)$.

Proof. We only prove (2) and (4), the rest is obvious.

(2) By the property of semi-algebra, there exist $A_1, \ldots, A_n \in \mathscr{S}$ mutually disjoint such that $B = A + A_1 + \cdots + A_n$. By the finite additivity and nonnegativeness of μ, we have $\mu(B) = \mu(A) + \sum_{i=1}^{n} \mu(A_i) \geqslant \mu(A)$.

(4) We prove first that Ω can be expressed as the union of countable many disjoint sets whose Φ-values are finite. Let $B_1 = A_1, B_n = A_n - \bigcup_{k=1}^{n-1} A_k \ \forall n \geqslant 1$. By the definition of semi-algebra, there exists $\exists B_{n1}, \ldots, B_{nk_n} \in \mathscr{S}$ mutually disjoint and $B_n = \sum_{i=1}^{k_n} B_{ni}$, so that

$$\Omega = \sum_{n=1}^{\infty} \sum_{i=1}^{k_n} B_{ni} \overset{\text{renumber}}{=\!=\!=\!=} \sum_{k=1}^{\infty} B'_k, \quad \{B'_k\} \subset \mathscr{S} \text{ mutually disjoint.}$$

From (3), it follows that $\Phi(A_n) < \infty$ $(\forall n \geqslant 1)$, which implies $\Phi(B_k) < \infty$ $(\forall k \geqslant 1)$. Then the desired assertion follows by letting $A'_n = A_n \cap B'_n$. □

Proposition 1.32.

(1) *(Finite subadditivity) Let μ be a finite additive measure on an algebra $\mathscr{F}$. For any $A, A_1, \ldots, A_n \in \mathscr{F}$ with $A \subset \bigcup_{k=1}^{n} A_k$, there holds $\mu(A) \leqslant \sum_{k=1}^{n} \mu(A_k)$.*

(2) *(σ-subadditivity) Let μ be a measure on an algebra $\mathscr{F}$. If $A \in \mathscr{F}$ and $\{A_n\}_{n \geqslant 1} \subset \mathscr{F}$ such that $A \subset \bigcup_{n=1}^{\infty} A_n$, then $\mu(A) \leqslant \sum_{n=1}^{\infty} \mu(A_n)$.*

Proof. (1) By induction, we only need to prove for $n = 2$. By the monotonicity and additivity,

$$\begin{aligned} \mu(A) &\leqslant \mu(A_1 \cup A_2) = \mu(A_1 + (A_2 - A_1)) \\ &= \mu(A_1) + \mu(A_2 - A_1) \leqslant \mu(A_1) + \mu(A_2). \end{aligned}$$

(2) Let $A_0 = \varnothing$. By the monotonicity and σ-additivity,

$$\mu(A) = \mu\left(\bigcup_{n=1}^{\infty} A_n \cap A\right) = \mu\left(\sum_{n=1}^{\infty} A \cap \left(A_n - \bigcup_{i\leqslant n-1} A_i\right)\right)$$

$$= \sum_{n=1}^{\infty} \mu\left(A \cap \left(A_n - \bigcup_{i\leqslant n-1} A_i\right)\right) \leqslant \sum_{n=1}^{\infty} \mu(A_n). \qquad \square$$

Definition 1.33. A function Φ on a class $\mathscr{C}$ is called lower continuous at $A \in \mathscr{C}$ if $\lim_{n\to\infty} \Phi(A_n) = \Phi(A)$ for any sequence $\mathscr{C} \ni A_n \uparrow A$, while it is called upper continuous at $A \in \mathscr{C}$ if $\lim_{n\to\infty} \Phi(A_n) = \Phi(A)$ for any sequence $\mathscr{C} \ni A_n \downarrow A$ with $\Phi(A_n) < \infty$ holds for some n. Moreover, Φ is called continuous at $A \in \mathscr{C}$ if it is both lower and upper continuous at A. Φ is called continuous if it is continuous at every $A \in \mathscr{C}$.

Note that we require the condition $\exists n$ such that $\Phi(A_n) < \infty$ for the upper continuity. Otherwise, the classical Lebesgue measure is excluded. More precisely, for $\Omega = \mathbb{R}$ and Φ being the Lebesgue measure, $A_n := (n, \infty)$ is decreasing to $\varnothing$, but $\Phi(\varnothing) = 0 \neq \infty = \lim_{n\to\infty} \Phi(A_n)$.

Theorem 1.34. *Let Φ be a signed measure on an algebra $\mathscr{F}$. Then Φ is continuous.*

Proof. Let $\mathscr{F} \ni A_n \uparrow A \in \mathscr{F}$. We have $A = \bigcup_{n=1}^{\infty} A_n = A_1 + \sum_{n=2}^{\infty} (A_n - A_{n-1})$. If there exists n such that $\Phi(A_n) = \infty$, then $\Phi(A) = \infty = \lim_{n\to\infty} \Phi(A_n)$. If $\Phi(A_n) < \infty$ for every n, then by the σ-additivity and subtractive property,

$$\Phi(A) = \Phi(A_1) + \sum_{n=2}^{\infty} \Phi(A_n - A_{n-1}) = \Phi(A_1) + \sum_{n=2}^{\infty} [\Phi(A_n) - \Phi(A_{n-1})]$$

$$= \Phi(A_1) + \lim_{n\to\infty} \sum_{k=2}^{n} [\Phi(A_k) - \Phi(A_{k-1})] = \lim_{n\to\infty} \Phi(A_n).$$

So, Φ is lower continuous.

On the other hand, let $\mathscr{F} \ni A_n \downarrow A \in \mathscr{F}$ with $\Phi(A_{n_0}) < \infty$ for some n_0. Then $A_{n_0} - A_n \uparrow A_{n_0} - A$, so that $\Phi(A_{n_0} - A_n) \to \Phi(A_{n_0} - A)$ by the lower continuity. This and the subtractive property imply $\Phi(A_n) \to \Phi(A)$. □

Corollary 1.35. *A measure on an algebra is continuous.*

The following theorem shows that when Φ is finitely additive, the continuity also implies the σ-additivity. This together with Theorem 1.34 implies the equivalence of the σ-additivity and the continuity of finitely additive functions on an algebra.

Theorem 1.36. *Let Φ be a finitely additive function on an algebra $\mathscr{F}$. If Φ satisfies one of the following conditions, then Φ is σ-additive:*

(a) *Φ is lower continuous;*
(b) *Φ is finite and is continuous at $\varnothing$.*

Proof. Let (a) hold. If $\{A_n\}_{n\geqslant 1} \subset \mathscr{F}$ mutually disjoint and $A = \sum\limits_{n=1}^{\infty} A_n \in \mathscr{F}$, then $B_n := \sum\limits_{k=1}^{n} A_k \uparrow A$. It follows from the lower continuity and the finite additivity that

$$\Phi(A) = \lim_{n\to\infty} \Phi(B_n) = \lim_{n\to\infty} \Phi\left(\sum_{k=1}^{n} A_k\right) = \lim_{n\to\infty} \sum_{k=1}^{n} \Phi(A_k) = \sum_{k=1}^{\infty} \Phi(A_k).$$

Let (b) hold, and let $\{A_n\}_{n\geqslant 1}$ and $\{B_n\}_{n\geqslant 1}$ as in above. We have $\mathscr{F} \ni A - B_n \downarrow \varnothing$. By the continuity at $\varnothing$ and the subtractive property, we obtain $0 = \lim\limits_{n\to\infty} \Phi(A - B_n) = \Phi(A) - \lim\limits_{n\to\infty} \Phi(B_n)$. Thus, $\Phi(A) = \sum\limits_{k=1}^{\infty} \Phi(A_k)$. □

1.2.2 Measure space

Definition 1.37. Let $\mathscr{A}$ be a σ-algebra in Ω and μ be a measure on $\mathscr{A}$. Then $(\Omega, \mathscr{A}, \mu)$ is called a measure space. If μ is a probability measure, then $(\Omega, \mathscr{A}, \mu)$ is called a probability space, and in this case, μ is often denoted by $\mathbb{P}$, and a measurable set is called an event.

Let $(\Omega, \mathscr{A}, \mathbb{P})$ be a probability space. By properties of a finite measure, we have the following assertions for the probability measure $\mathbb{P}$:

(1) (*Nonnegativity*) $\mathbb{P}(A) \geqslant 0, \ \forall A \in \mathscr{A}$.
(2) (*Normality*) $\mathbb{P}(\Omega) = 1$.
(3) (*σ-additivity, hence finite additivity*) If $A_n \in \mathscr{A}, n = 1, 2, \ldots$ are mutually disjoint, then

$$\mathbb{P}\left(\sum_{n=1}^{\infty} A_n\right) = \sum_{n=1}^{\infty} \mathbb{P}(A_n).$$

(4) (*Subtractive property, hence monotonicity*) If $A \subset B, A, B \in \mathscr{A}$, then

$$\mathbb{P}(B - A) = \mathbb{P}(B) - \mathbb{P}(A) \Rightarrow \mathbb{P}(B) \geqslant \mathbb{P}(A).$$

(5) (*Additive formula*) $\mathbb{P}(A \cup B) = \mathbb{P}(A) + \mathbb{P}(B) - \mathbb{P}(A \cap B)$. In general, $\forall \{A_n\}_{n=1}^{\infty} \subset \mathscr{A}$, we have

$$\begin{aligned}
&\mathbb{P}\left(\bigcup_{k=1}^{n} A_k\right) \\
&= \sum_{k=1}^{n} \mathbb{P}(A_k) - \sum_{1 \leqslant i < j \leqslant n} \mathbb{P}(A_i \cap A_j) + \cdots + (-1)^{n-1} \mathbb{P}(A_1 \cap \cdots \cap A_n) \\
&= \sum_{\ell=1}^{n} (-1)^{\ell-1} \sum_{1 \leqslant i_1 < i_2 < \cdots < i_\ell \leqslant n} \mathbb{P}(A_{i_1} \cdots A_{i_\ell}).
\end{aligned}$$

(6) (*Continuity*) For $A, A_n \in \mathscr{A}, n \geqslant 1$,

$$A_n \uparrow A \Rightarrow \mathbb{P}(A_n) \uparrow \mathbb{P}(A); \quad A_n \downarrow A \Rightarrow \mathbb{P}(A_n) \downarrow \mathbb{P}(A).$$

Example 1.38 (Geometric probability model). Let $\Omega \subset \mathbb{R}$ be a Lebesgue measurable set and $0 < |\Omega| < \infty$, where $|\cdot|$ denotes the Lebesgue measure. Assume $\mathscr{A}$ is a class of Lebesgue measurable subsets of Ω, $\mathbb{P}(A) = \frac{|A|}{|\Omega|}, A \in \mathscr{A}$. Then $(\Omega, \mathscr{A}, \mathbb{P})$ is a probability space.

1.3 Extension and Completion of Measure

As explained in the beginning of this chapter, a measure is often easily defined on a semi-algebra. So, to build up a measure space, it is crucial to extend a measure from a semi-algebra to the induced

σ-algebra. In this section, we first extend a measure from a semi-algebra to its generated algebra, which is easy to do according to the formula of the induced algebra, and then further extend to the generated σ-algebra and finally make completion of the resulting measure space.

1.3.1 Extension from semi-algebra to the induced algebra

Definition 1.39. Let $\mathscr{C}_1 \subset \mathscr{C}_2$ be two classes of sets in Ω, and let μ_i be measures (or finitely additive measures) defined on $\mathscr{C}_i\,(i = 1, 2)$, respectively. If $\mu_1(A) = \mu_2(A)$ holds for any $A \in \mathscr{C}_1$, then we call μ_2 an extension of μ_1 from $\mathscr{C}_1$ to $\mathscr{C}_2$ and call μ_1 the restriction of μ_2 on $\mathscr{C}_1$ which is denoted by $\mu_1 = \mu_2|_{\mathscr{C}_1}$.

Theorem 1.40. *Let μ be a measure (or finitely additive measure) on a semi-algebra $\mathscr{S}$. Then μ can be uniquely extended to a measure (or finitely additive measure) $\widetilde{\mu}$ on $\mathscr{F}(\mathscr{S})$.*

Proof. By Theorem 1.7, for any $A \in \mathscr{F}(\mathscr{S})$, there exist $B_1, \ldots, B_n \in \mathscr{S}$ mutually disjoint such that $A = \sum_{i=1}^{n} B_i$. Define $\widetilde{\mu}(A) = \sum_{i=1}^{n} \mu(B_i)$. First, we prove that $\widetilde{\mu}(A)$ is independent of the choices of $\{B_i\}$. Let $B'_1, \ldots, B'_{n'} \in \mathscr{S}$ be mutually disjoint such that $A = \sum_{i=1}^{n'} B'_i$. Then $B'_i = \sum_{j=1}^{n} B'_i \cap B_j$. Since $B'_i \cap B_j \in \mathscr{S}$, by the finite additivity, we have $\mu(B'_i) = \sum_{j=1}^{n} \mu(B'_i \cap B_j)$. So,

$$\sum_{i=1}^{n'} \mu(B'_i) = \sum_{i=1}^{n'} \sum_{j=1}^{n} \mu(B'_i \cap B_j) = \sum_{j=1}^{n} \sum_{i=1}^{n'} \mu(B'_i \cap B_j)$$
$$= \sum_{j=1}^{n} \mu(B_j) = \widetilde{\mu}(A).$$

Thus, $\widetilde{\mu}(A)$ is independent of the choice of $\{B_i\}$.

Next, we prove that $\widetilde{\mu}$ is a measure (or finitely additive measure). It is obvious for nonnegativeness and uniqueness, as well as finite additivity. We are going to prove the σ-additivity. Let $\{A_n\}_{n\geqslant 1} \subset \mathscr{F}(\mathscr{S})$ be mutually disjoint such that $A = \sum_{n=1}^{\infty} A_n \in \mathscr{F}(\mathscr{S})$. Take $B_1, \dots, B_k \in \mathscr{S}$ mutually disjoint such that $A = \sum_{i=1}^{k} B_i$. Again, $\forall n \geqslant 1$, take $C_{n1}, \dots, C_{nk_n} \in \mathscr{S}$ mutually disjoint, satisfying $A_n = \sum_{i=1}^{k_n} C_{ni}$. Then $\forall i \leqslant k$, $B_i = \sum_{n=1}^{\infty} A_n \cap B_i = \sum_{n=1}^{\infty} \sum_{l=1}^{k_n} B_i \cap C_{nl}$ is the union of mutually disjoint subsets in $\mathscr{S}$. By the σ-additivity of μ, we have $\mu(B_i) = \sum_{j=1}^{\infty} \sum_{l=1}^{k_j} \mu(B_i \cap C_{jl})$. From this and the finite additivity, it follows that

$$\begin{aligned}
\tilde{\mu}(A) = \tilde{\mu}\left(\sum_{i=1}^{k} B_i\right) &= \sum_{i=1}^{k} \sum_{n=1}^{\infty} \sum_{l=1}^{k_n} \mu\left(B_i \cap C_{nl}\right) \\
&= \sum_{n=1}^{\infty} \sum_{l=1}^{k_n} \sum_{i=1}^{k} \mu\left(B_i \cap C_{nl}\right) = \sum_{n=1}^{\infty} \tilde{\mu}\left(A_n\right).
\end{aligned}$$
□

By applying Proposition 1.32 to $\tilde{\mu}$ on $\mathscr{F}(\mathscr{C})$, we obtain the following result.

Corollary 1.41. *Let μ ba a finite additive measure on a semi-algebra $\mathscr{S}$, and let $A, A_1, \dots, A_n \in \mathscr{S}$.*

(a) *If $A_1, \dots, A_n$ are mutually disjoint and $\sum_{i=1}^{n} A_i \subset A$, then $\sum_{i=1}^{n} \mu(A_i) \leqslant \mu(A)$.*

(b) *If $\bigcup_{i=1}^{n} A_i \supset A$, then $\sum_{i=1}^{n} \mu(A_i) \geqslant \mu(A)$.*

If μ is σ-additive, the above assertions hold for $n = \infty$.

1.3.2 Extension from semi-algebra to the generated σ-algebra

Theorem 1.42 (Measure extension theorem). *Let μ be a measure on a semi-algebra $\mathscr{S}$ in Ω. Then it can be extended to a measure on $\sigma(\mathscr{S})$. If furthermore μ is σ-finite, then the extension is unique.*

Following the line of the Lebesgue measure theory, we first define an outer measure for every subset of Ω by the covering procedure and then prove that the restriction of the outer measure is σ-additive on the generated σ-algebra.

Definition 1.43. Let μ be a measure on a semi-algebra $\mathscr{S}$ in Ω. For any $A \subset \Omega$,

$$\mu^*(A) := \inf \left\{ \sum_{n=1}^{\infty} \mu(A_n) : A \subset \bigcup_{n=1}^{\infty} A_n, A_n \in \mathscr{S} \right\}$$

is called the outer measure of A, and the function μ^* defined on the largest σ-algebra 2^{Ω} is called the outer measure generated by μ.

Property 1.44.

(1) $\mu^*|_{\mathscr{S}} = \mu$.
(2) $\mu^*(A) \leqslant \mu^*(B), \forall A \subset B$.
(3) $\mu^*\left(\bigcup_{n=1}^{\infty} A_n\right) \leqslant \sum_{n=1}^{\infty} \mu^*(A_n),\ \forall A_n \subset \Omega, n \geqslant 1$.

Proof. (1) As $A \subset A$, by letting $A_1 = A, A_n = \varnothing, n \geqslant 2$, we have $\mu^*(A) \leqslant \mu(A)$. On the other hand, by the sub σ-additivity of μ, it follows that $\mu(A) \leqslant \sum_{n=1}^{\infty} \mu(A_n)$ for any sequence $\{A_n\} \subset \mathscr{S}$ with $\bigcup_{n=1}^{\infty} A_n \supset A$. So, $\mu^*(A) \geqslant \mu(A)$.

(2) Obvious.
(3) For any $\varepsilon > 0$ and $n \geqslant 1$, take $A_{n1}, A_{n2}, \dots \in \mathscr{S}$ such that $\bigcup_{i=1}^{\infty} A_{ni} \supset A_n$ and $\mu^*(A_n) \geqslant \sum_{i=1}^{\infty} \mu(A_{ni}) - \varepsilon/2^n$. Thus, $\bigcup_{n=1}^{\infty} \bigcup_{i=1}^{\infty} A_{ni} \supset \bigcup_{n=1}^{\infty} A_n$, and by the definition of μ^*

$$\mu^*\left(\bigcup_{n=1}^{\infty} A_n\right) \leqslant \sum_{n=1}^{\infty}\sum_{i=1}^{\infty} \mu(A_{ni}) \leqslant \sum_{n=1}^{\infty}\left(\mu^*(A_n) + \frac{\varepsilon}{2^n}\right)$$
$$= \sum_{n=1}^{\infty} \mu^*(A_n) + \varepsilon.$$

Let $\varepsilon \downarrow 0$ to derive the assertion.

□

If μ^* were a measure on 2^Ω, then the restriction $\mu^*|_{\sigma(\mathscr{S})}$ would be an extended measure as desired. However, this is in general not true as the Lebesgue measure is already a counterexample. So, we need to find a class $\mathscr{A}^*$ of "regular" sets such that $\mathscr{A}^* \supset \sigma(\mathscr{S})$ and μ^* is σ-additive on $\mathscr{A}^*$. An intuition to select a "regular" set is that it does not leads to any loss of outer measures when using the set to cut others. In this spirit, we introduce the notion of μ^*-measurable set as follows.

Definition 1.45. A set $A \subset \Omega$ is called μ^*-measurable if

$$\mu^*(D) = \mu^*(A \cap D) + \mu^*(A^c \cap D), \quad \forall D \subset \Omega.$$

Let $\mathscr{A}^* = \{A \subset \Omega : A \text{ is } \mu^*\text{-measurable}\}$.

We shall prove that $\mathscr{A}^*$ is a σ-algebra including $\mathscr{S}$ and μ^* is a measure on $\mathscr{A}^*$. For this, we first study the properties of μ^* and $\mathscr{A}^*$. The following is a consequence of Property 1.44(3).

Property 1.46. A is a μ^* measurable set if and only if

$$\mu^*(D) \geqslant \mu^*(A \cap D) + \mu^*(A^c \cap D), \quad \forall\, D \subset \Omega.$$

Property 1.47. $\mathscr{A}^* \supset \mathscr{S}$.

Proof. Let $A \in \mathscr{S}, D \subset \Omega$. For any $\varepsilon > 0$, take $\{A_n\} \subset \mathscr{S}$ such that $\sum_{n=1}^{\infty} A_n \supset D, \mu^*(D) \geqslant \sum_{n=1}^{\infty} \mu(A_n) - \varepsilon$. Then by σ-subadditivity of μ^* and finite additivity of μ on $\mathscr{F}(\mathscr{S})$, it follows that

$$\begin{aligned}\mu^*(A^c \cap D) + \mu^*(A \cap D) &\leqslant \sum_{n=1}^{\infty} \left[\mu\left(A_n \cap A\right) + \mu\left(A^c \cap A_n\right)\right] \\ &= \sum_{n=1}^{\infty} \mu(A_n) \leqslant \mu^*(D) + \varepsilon.\end{aligned}$$

Let $\varepsilon \downarrow 0$ to derive $\mu^*(D) \geqslant \mu^*(A \cap D) + \mu^*(A^c \cap D)$. Thus, $A \in \mathscr{A}^*$ by Property 1.46. □

Theorem 1.48.

(1) *$\mathscr{A}^*$ is a σ-algebra, so that $\mathscr{A}^* \supset \sigma(\mathscr{S})$.*

(2) *If* $\{A_n\} \subset \mathscr{A}^*$ *are mutually disjoint and* $A = \sum_{n=1}^{\infty} A_n$, *then* $\forall D \subset \Omega$,

$$\mu^*(D \cap A) = \sum_{n=1}^{\infty} \mu^*(D \cap A_n).$$

(3) *The restriction of* μ^* *on* $\mathscr{A}^*$ *is a measure on* $\mathscr{A}^*$.

Proof. (1) We first prove that $\mathscr{A}^*$ is an algebra. Since $\mathscr{A}^* \supset \mathscr{S}$, $\varnothing, \Omega \in \mathscr{A}^*$. It is obvious that $A \in \mathscr{A}^*$ implies $A^c \in \mathscr{A}^*$. So, it suffices to prove that $A, B \in \mathscr{A}^* \Rightarrow A \cap B \in \mathscr{A}^*$. By subadditivity,

$$\begin{aligned}
\mu^*(D) &= \mu^*(A \cap D) + \mu^*(A^c \cap D) \\
&= \mu^*(A \cap B \cap D) + \mu^*(A \cap B^c \cap D) + \mu^*(A^c \cap D) \\
&\geqslant \mu^*(A \cap B \cap D) + \mu^*((A^c \cup B^c) \cap D).
\end{aligned}$$

Thus, $A \cap B \in \mathscr{A}^*$ by Property 1.46.

Next, we prove that $\mathscr{A}^*$ is a monotone class. Let $A_n \in \mathscr{A}^*$ such that $A_n \uparrow A$. By Property 1.44(2) and letting $A_0 = \varnothing$, we have

$$\begin{aligned}
\mu^*(D) &= \mu^*(A_1 \cap D) + \mu^*(A_1^c \cap D) \\
&= \mu^*(A_1 \cap D) + \mu^*(A_2 \cap A_1^c \cap D) + \mu^*(D \cap A_2^c) \\
&= \cdots = \sum_{i=1}^{n} \mu^*((A_i - A_{i-1}) \cap D) + \mu^*(D \cap A_n^c) \\
&\geqslant \sum_{i=1}^{n} \mu^*((A_i - A_{i-1}) \cap D) + \mu^*(D \cap A^c). \qquad (1.1)
\end{aligned}$$

Letting $n \to \infty$, we derive

$$\mu^*(D) \geqslant \mu^*(D \cap A^c) + \sum_{i=1}^{\infty} \mu^*(D \cap (A_{i-1} - A_i)) \geqslant \mu^*(D \cap A^c) + \mu^*(D \cap A).$$

Thus, $A \in \mathscr{A}^*$.

Therefore, $\mathscr{A}^*$ is a σ-algebra by the monotone class theorem.

(2) Let $A = \sum_{n=1}^{\infty} A_n$ with $A_n \in \mathscr{A}^*$ mutually disjoint. Then $A \in \mathscr{A}^*$. By Property 1.44-(2), it suffices to prove $\mu^*(D \cap A) \geqslant \sum_{n=1}^{\infty} \mu^*$

$(D \cap A_n)$. Replacing D by $A \cap D$ and A_n by $\sum_{i=1}^{n} A_i$ in (1.1), we obtain $\mu^*(D \cap A) \geqslant \sum_{i=1}^{n} \mu^*(D \cap A_i)$. Then the proof is finished by letting $n \uparrow \infty$.

(3) The σ-additivity of μ^* on $\mathscr{A}^*$ is obtained by letting $D = \Omega$ in (2). □

Proof of Theorem 1.42. Since $\mathscr{A}^* \supset \sigma(\mathscr{S})$, the restriction of μ^* on $\sigma(\mathscr{S})$ is obviously a measure, and $\mu^*(A) = \mu(A)$ for $A \in \mathscr{S}$. Thus, there exists an extension of μ on $\sigma(\mathscr{S})$.

Now, let μ be σ-finite on $\mathscr{S}$. By Property 1.31(4), there exist mutually disjoint $\{A_n\} \subset \mathscr{S}$ such that $\Omega = \sum_{n=1}^{\infty} A_n$ and $\mu(A_n) < \infty$, $n \geqslant 1$. If both μ_1 and μ_2 are measures on $\sigma(\mathscr{S})$ extended from μ, it suffices to prove $\mu_1(A \cap A_n) = \mu_2(A \cap A_n)$ for $A \in \sigma(\mathscr{S})$ and $n \geqslant 1$. For this, let $\mathscr{M}_n = \{A : A \in \sigma(\mathscr{S}), \mu_1(A \cap A_n) = \mu_2(A \cap A_n)\}$. Then $\mathscr{M}_n \supset \mathscr{S}$. By the unique extension of μ on $\mathscr{F}(\mathscr{S})$, we have $\mathscr{M}_n \supset \mathscr{F}(\mathscr{S})$, thus by the monotone class theorem, it is sufficient to show that $\mathscr{M}$ is a monotone class, which can be derived by the continuity of measures. □

Corollary 1.49. *If $\mathscr{S}$ is a semi-algebra in Ω, and $\mathbb{P}$ is a probability measure on $\mathscr{S}$, then $\mathbb{P}$ can be uniquely extended to a probability measure on $\sigma(\mathscr{S})$.*

1.3.3 Completion of measures

Definition 1.50. Let $(\Omega, \mathscr{A}, \mu)$ be a measure space. A subset B of Ω is called a μ-null set if there exists $A \in \mathscr{A}$ such that $B \subset A$ and $\mu(A) = 0$. If all μ-null sets are contained in $\mathscr{A}$, then $(\Omega, \mathscr{A}, \mu)$ is called a complete measure space.

Theorem 1.51. *For a measure space $(\Omega, \mathscr{A}, \mu)$, let*

$$\bar{\mathscr{A}} = \{A \cup N : A \in \mathscr{A}, N \text{ is a } \mu\text{-null set}\}$$

and define $\bar{\mu}(A \cup N) := \mu(A)$ for $A \in \mathscr{A}$ and N a μ-null set. Then $(\Omega, \bar{\mathscr{A}}, \bar{\mu})$ is a complete measure space, which is called the completion of $(\Omega, \mathscr{A}, \mu)$.

Proof. We first prove that $\bar{\mathscr{A}}$ is a σ-algebra. By the σ-subadditivity of μ, the union of countable many μ-null sets is still μ-null, so that $\bar{\mathscr{A}}$ is closed under countable union. It remains to be proved that $\bar{\mathscr{A}}$ is closed under complement. Let $A \cup N \in \bar{\mathscr{A}}$ with $A \in \mathscr{A}$ and N a μ-null set. Assume $B \in \mathscr{A}$ such that $B \supset N$ and $\mu(B) = 0$. Then $(A \cup N)^c = A^c \cap N^c = A^c \cap B^c + A^c \cap (N^c - B^c)$. As $A^c \cap (N^c - B^c) \subset \Omega - B^c = B$, and $\mu(B) = 0$, $A^c \cap (N^c - B^c)$ is μ-null. Moreover, $A^c \cap B^c \in \mathscr{A}$, so that $(A \cup N)^c \in \bar{\mathscr{A}}$ by definition.

Next, it is easy to check that $\bar{\mu}$ is σ-additive on $\bar{\mathscr{A}}$. It suffices to prove that $(\Omega, \bar{\mathscr{A}}, \bar{\mu})$ is complete. Let $\bar{N}$ be a $\bar{\mu}$-null set. Then $\exists \bar{B} \in \bar{\mathscr{A}}$ such that $\bar{\mu}(\bar{B}) = 0$ and $\bar{B} \supset \bar{N}$. By $\bar{B} \in \bar{\mathscr{A}}$, we have $\bar{B} = A \cup N$ for some $A \in \mathscr{A}$ and a μ-null set N. Then $0 = \bar{\mu}(\bar{B}) = \mu(A)$. Take $B \in \mathscr{A}, B \supset N$ such that $\mu(B) = 0$. We have $\bar{N} \subset \bar{B} \subset A \cup B$, and $\mu(A \cup B) = 0$. So, $\bar{N}$ is μ-null, and hence, $\bar{N} \in \bar{\mathscr{A}}$ by the definition of $\bar{\mathscr{A}}$. □

Theorem 1.52. *Let μ be a measure on a semi-algebra $\mathscr{S}$, and let μ^* be the induced outer measure. Then for any $A \subset \Omega$ with $\mu^*(A) < \infty$, $\exists B \in \sigma(\mathscr{S})$ such that*

(i) $A \subset B$,
(ii) $\mu^*(A) = \mu(B)$,
(iii) $\mu^*(C) = 0, \forall C \subset B - A$ *and* $C \in \sigma(\mathscr{S})$.

The above B is called a measurable cover of A.

Proof. $\forall n \geqslant 1$, take $\{F_{n_k}\}_{k \geqslant 1} \subset \mathscr{S}$ such that $A \subset \bigcup_{k=1}^{\infty} F_{n_k}$ and $\sum_{k=1}^{\infty} \mu(F_{n_k}) \leqslant \mu^*(A) + 1/n$. Let $B_n = \bigcup_{k=1}^{\infty} F_{n_k}$. Then $\mu^*(A) \leqslant \mu^*(B_n)$. By setting $B = \bigcap_{n=1}^{\infty} B_n$, we have $B \in \sigma(\mathscr{S})$ with $B \supset A$ and $\mu^*(B) = \mu^*(A)$. If $C \in \sigma(\mathscr{S})$ and $C \subset B - A$, then $A \subset B - C$. Thus, $\mu^*(A) \leqslant \mu^*(B - C) = \mu(B) - \mu(C)$. From this and $\mu^*(B) = \mu^*(A) < \infty$, it follows that $\mu^*(C) = 0$. □

Theorem 1.53. *Let μ be a σ-finite measure on a semi-algebra $\mathscr{S}$, and let μ^* be the induced outer measure. Then $(\Omega, \mathscr{A}^*, \mu^*)$ is the completion of $(\Omega, \sigma(\mathscr{S}), \mu)$.*

Proof. By Theorem 1.51, we only need to prove $\mathscr{A}^* = \bar{\mathscr{A}}$.

Let $\bar{A} \in \bar{\mathscr{A}}$. Then $\exists A \in \sigma(\mathscr{S})$ and a μ-null set N such that $\bar{A} = A \cup N$. It is clear $\mathscr{A}^*$ contains all μ null sets, so that $\bar{A} \in \mathscr{A}^*$.

Conversely, for $A \in \mathscr{A}^*$ with $\mu^*(A) < \infty$, let B be a measurable cover of A and C be a measurable cover of $B - A$. Then $A = (B-C) \cup (C - (B-A))$, where $B - C \in \sigma(\mathscr{S}), C - (B-A)$ are μ-null. Thus, $A \in \bar{\mathscr{A}}$. When $\mu^*(A) = \infty$, by the σ-finiteness of μ, we have $\exists \{A_n\}_{n\geqslant 1} \subset \mathscr{S}$ such that $\sum\limits_{n=1}^{\infty} A_n = \Omega$ and $\mu(A_n) < \infty, n \geqslant 1$. From the previous proofs, it follows $A \cap A_n \in \bar{\mathscr{A}}$ for $n \geqslant 1$. Thus, $A = \sum\limits_{n=1}^{\infty} A_n \cap A \in \bar{\mathscr{A}}$. □

Theorem 1.54. *Let μ be a σ-finite measure on a semi-algebra $\mathscr{S}$. Then μ has a unique measure extension on $\mathscr{A}^*$.*

Proof. By Theorem 1.42, μ is uniquely extended to a measure on $\sigma(\mathscr{S})$, and the extension is the restriction of μ^* on $\sigma(\mathscr{S})$. Let μ_1 be another measure on $\mathscr{A}^*$ which extends μ. Then for any $A \cup N \in \mathscr{A}^* = \bar{\mathscr{A}}$, where $A \in \sigma(\mathscr{S})$ and N is μ-null, we have

$$\begin{aligned} \mu^*(A \cup N) &= \mu^*(A) = \mu_1(A) \leqslant \mu_1(A \cup N) \leqslant \mu_1(A) + \mu_1(N) \\ &= \mu^*(A) + \mu_1(N) = \mu^*(A \cup N) + \mu_1(N). \end{aligned}$$

Let $B \in \sigma(\mathscr{S})$ such that $B \supset N$ and $\mu(B) = 0$. We obtain $\mu_1(N) \leqslant \mu_1(B) = \mu(B) = 0$. So, $\mu^*(A \cup N) = \mu_1(A \cup N)$. □

Theorem 1.55. *Let μ be a measure on a semi-algebra $\mathscr{S}$ in Ω. Then $\forall A \in \mathscr{A}^*$ with $\mu^*(A) < \infty$ and $\forall \varepsilon > 0$, there exists $A_\varepsilon \in \mathscr{F}(\mathscr{S})$ such that $\mu^*(A \Delta A_\varepsilon) < \varepsilon$, where $A \Delta A_\varepsilon) := (A - A_\varepsilon) + (A_\varepsilon - A)$.*

Proof. $\forall \varepsilon > 0$, there exists a sequence $\{B_n\}_{n\geqslant 1} \subset \mathscr{S}$ such that $\bigcup\limits_{n=1}^{\infty} B_n \supset A$ and $\mu^*(A) \leqslant \sum\limits_{n=1}^{\infty} \mu^*(B_n) \leqslant \mu^*(A) + \frac{\varepsilon}{2}$. Since $\mu^*(A) < \infty$, $\sum\limits_{n=1}^{\infty} \mu^*(B_n) < \infty$. Take $n_0 \geqslant 1$ such that $\sum\limits_{n>n_0} \mu^*(B_n) < \frac{\varepsilon}{2}$. Let $A_\varepsilon = \sum\limits_{n=1}^{n_0} B_n$ and $B_\varepsilon = \sum\limits_{n>n_0} B_n$. Then $A_\varepsilon \in \mathscr{F}(\mathscr{S})$. By the σ-subadditivity of μ^*, it follows that $\mu^*(B_\varepsilon) < \frac{\varepsilon}{2}$. So, $\mu^*((A_\varepsilon \cup B_\varepsilon) - A) < \frac{\varepsilon}{2}$ by monotone property. As $A - A_\varepsilon \subset B_\varepsilon$ and $A_\varepsilon - A \subset (A_\varepsilon \cup B_\varepsilon) - A$, we obtain $\mu^*(A \Delta A_\varepsilon) = \mu^*((A - A_\varepsilon) + (A_\varepsilon - A)) < \varepsilon$. □

1.4 Exercises

1. Prove Proposition 1.6.

2. Prove Property 1.10.

3. Let $\mathscr{C}$ be a class of subsets of Ω. Then $\forall A \in \sigma(\mathscr{C})$, there exists a countable sub-class $\mathscr{C}_A$ of $\mathscr{C}$ such that $A \in \sigma(\mathscr{C}_1)$.

4. (*Countable generation*) A σ-algebra $\mathscr{A}$ is called countably generated if there exists a countable sub-class $\mathscr{C}$ such that $\sigma(\mathscr{C}) = \mathscr{A}$. Prove that the Borel σ-algebra $\mathscr{B}^d$ in $\mathbb{R}^d$ is countably generated.

5. Let $\{\mathscr{C}_n\}_{n\geqslant 1}$ be an increasing sequence for classes of sets in Ω.

   (a) If $\{\mathscr{C}_n\}_{n\geqslant 1}$ are algebras, prove that $\bigcup\limits_{n=1}^{\infty} \mathscr{C}_n$ is an algebra.

   (b) Exemplify that $\bigcup\limits_{n=1}^{\infty} \mathscr{C}_n$ is not a σ-algebra, but $\{\mathscr{C}_n\}_{n\geqslant 1}$ are σ-algebras.

6. Prove Theorem 1.19.

7. Prove that a σ-algebra is either finite or uncountable.

8. Let $(\Omega_i, A_i), 1 \leqslant i \leqslant n$, be measure spaces. Prove that

$$\mathscr{C} := \{A_1 \times \cdots \times A_n : A_i \in \mathscr{A}_i\}$$

   is a semi-algebra in $\Omega := \Omega_1 \times \cdots \times \Omega_n$.

9. Prove Theorem 1.27.

10. Exemplify that an additive measure on a class of sets may not be finitely additive.

11. Exemplify that the σ-algebra generated by a semi-algebra $\mathscr{S}$ cannot be expressed as

$$\sigma(\mathscr{S}) = \left\{ \sum_{n=1}^{\infty} A_n : \forall n \geqslant 1, A_n \in \mathscr{S} \right\},$$

    and prove this formula when Ω is finite or countable.

12. Let $(\Omega_n, \mathscr{A}_n, \mu_n), n \geqslant 1$, be a sequence of measure spaces with $\{\Omega_n\}$ mutually disjoint. Set

$$\Omega = \sum_{n=1}^{\infty} \Omega_n, \qquad \mathscr{A} = \{A \subset \Omega : \forall n \geqslant 1, A \cap \Omega_n \in \mathscr{A}_n\},$$

$$\mu(A) = \sum_{n=1}^{\infty} \mu_n(A \cap \Omega_n), \quad A \in \mathscr{A}.$$

Prove that $(\Omega, \mathscr{A}, \mu)$ is a measure space.

13. Let Ω be infinite, and let $\mathscr{F}$ be the class of finite subsets of Ω and their complements. Define $\mathbb{P}(A) = 0$ if A is finite and $= 1$ if A^c is finite.
   (a) Prove that $\mathscr{F}$ is an algebra and $\mathbb{P}$ is finitely additive.
   (b) When Ω is countable, prove that $\mathbb{P}$ is not σ-additive.
   (c) When Ω is uncountable, prove that $\mathbb{P}$ is σ-additive.

14. Prove Proposition 1.30.

15. Let $(\Omega, \mathscr{A}, \mathbb{P})$ be a probability measure space without atom, i.e. for any $A \in \mathscr{A}$ with $\mathbb{P}(A) > 0$, there exists $B \in \mathscr{A}$ such that $B \subset A$ and $0 < \mathbb{P}(B) < \mathbb{P}(A)$. For any $A \in \mathscr{A}$ with $\mathbb{P}(A) > 0$, prove that $\{\mathbb{P}(B) : B \in \mathscr{A}, B \subset A\} = [0, \mathbb{P}(A)]$.

16. Prove Corollary 1.35.

17. Let $([0,1], \mathscr{B}([0,1]), \mu)$ be a finite measure space with $\mu(\{x\}) = 0, \forall x \in [0,1]$. $\forall \varepsilon > 0$. Prove
   (a) $\forall x \in [0,1]$, there exists an interval $I \ni x$ such that $\mu(I) \leqslant \varepsilon$;
   (b) there exists a dense subset A of $[0,1]$ such that $\mu(A) \leqslant \varepsilon$.

18. Prove Corollary 1.41.

19. Prove Property 1.46.

20. Construct an example of a measure μ on a semi-algebra $\mathscr{C}$ such that it has more than one extended measures on $\sigma(\mathscr{C})$.

21. Prove that a measure space $(\Omega, \mathscr{A}, \mu)$ is complete if and only if $\mathscr{A} \supset \{A \subset \Omega : \mu^*(A) = 0\}$.

22. Let μ be a finite measure on a semi-algebra $\mathscr{S}$. Let

$$\mu_*(A) = \sup\left\{\sum_n \mu(A_n) : A_n \in \mathscr{S} \text{ mutually disjoint}, \sum_n A_n \subset A\right\},$$

$$\mathscr{A}_* = \{A \subset \Omega : \mu^*(A) = \mu_*(A)\}.$$

Prove $\mathscr{A}^* \supset \mathscr{A}_*$.

23. Let $(\Omega, \mathscr{A}, \mu)$ be a measure space. Prove that $N \subset \Omega$ is μ-null if and only if $\mu^*(N) = 0$.

24. For a measure space $(\Omega, \mathscr{A}, \mu)$, let $A_i, B_i \subset \Omega$ satisfy $\mu^*(A_i \Delta B_i) = 0, i \geqslant 1$. Prove that

$$\mu^*\left(\sum_{i=1}^{\infty} A_i\right) = \mu^*\left(\sum_{i=1}^{\infty} B_i\right).$$

25. Let $\mathscr{C} = \{C_{a,b} = [-b, -a) \cup (a, b] : 0 < a < b\}$ and define $\mu(C_{a,b}) = b - a$. Prove that μ can be extended to a measure on $\sigma(\mathscr{C})$. Ask whether $[1, 2]$ is μ^*-measurable.

26. Let $f : [0, \infty) \to [0, \infty)$ be strictly increasing, strictly convex and $f(0) = 0$. $\forall A \subset (0, 1]$, define $\mu^*(A) = f(\lambda^*(A))$, where λ^* is the Lebesgue outer measure. Prove that μ^* satisfies $\mu^*(\varnothing) = 0$, nonnegativeness, monotonicity and σ-subadditivity.

27. Let $(\Omega, \mathscr{A}, \mathbb{P})$ be a probability space, and let $\Omega \supset A \notin \mathscr{A}$. Prove that $\mathbb{P}$ can be extended to a probability measure on $\mathscr{A}_1 := \sigma(\mathscr{A} \cup \{A\})$.

28. Let $f : \mathbb{R} \ni x \longmapsto \frac{x}{3} \in \mathbb{R}$ and $A_0 = [0, 1]$. Prove that $A_{n+1} = f(A_n) \cup (\frac{2}{3} + f(A_n))$ $(n \geqslant 0)$ is decreasing in $n \geqslant 0$, where $f(A_n) := \{f(x) : x \in A_n\}$. The limit of A_n is denoted by C, which is called the Cantor set. Prove that the Lebesgue measure of C is 0.

Chapter 2

Random Variable and Measurable Function

Given a probability space $(\Omega, \mathscr{A}, \mathbb{P})$, we define random variables and their distribution functions as follows.

Definition 2.1.

(1) A real function $\xi : \Omega \to \mathbb{R}$ is called a random variable on $(\Omega, \mathscr{A}, \mathbb{P})$ if $\{\omega : \xi(\omega) < x\} \in \mathscr{A}$ for every $x \in \mathbb{R}$. Let $i = \sqrt{-1}$. We call $\xi = \eta + i\zeta$ a complex random variable on $(\Omega, \mathscr{A}, \mathbb{P})$ provided η and ζ are random variables.

(2) If $\xi_1, \dots, \xi_n$ are real (complex) random variables on $(\Omega, \mathscr{A}, \mathbb{P})$, then vector-valued function $\xi := (\xi_1, \dots, \xi_n)$ is called an n-dimensional real (complex) random variable on $(\Omega, \mathscr{A}, \mathbb{P})$. A multi-dimensional random variable is also called random vector.

(3) The distribution function of a random variable $\xi := (\xi_1, \dots, \xi_n)$ is defined as

$$F : \mathbb{R}^n \ni (x_1, x_2, \dots, x_n) \mapsto \mathbb{P}(\xi_i < x_i : 1 \leqslant i \leqslant n).$$

(4) Let $\xi := (\xi_1, \dots, \xi_n)$ and $\eta := (\eta_1, \dots, \eta_n)$ be two random variables on $(\Omega, \mathscr{A}, \mathbb{P})$. If

$$\mathbb{P}(\xi_i \neq \eta_i) = 0, \quad 1 \leqslant i \leqslant n,$$

then these two random variables are called identical almost surely, denoted by $\xi = \eta$, a.s. If they have the same distribution function, we call them identically distributed.

In this chapter, we first extend the concept of random variables to measurable functions on a measurable space and then study the construction of measurable functions and convergence theorems. These provide a basis for the following chapter to define and study the integral (in particular, expectation) of measurable functions (random variables).

2.1 Measurable Function

2.1.1 Definition and properties

Let $\mathscr{B}$ be the Borel σ-algebra of $\mathbb{R}$ and $\bar{\mathbb{R}} = [-\infty, \infty], \bar{\mathscr{B}} = \sigma(\mathscr{B} \cup \{\infty\} \cup \{-\infty\})$. Let $\bar{\mathbb{R}}^n$ be the n-dimensional product space of $\bar{\mathbb{R}}$ and $\bar{\mathscr{B}}^n$ be the product σ-algebra. Similarly, we can define n-dimensional product space $\bar{\mathbb{C}}^n$ of the generalized complex plane $\bar{\mathbb{C}}$ and the product σ-algebra $\bar{\mathscr{B}}_c^n$.

Definition 2.2. Let $(\Omega, \mathscr{A})$ and $(E, \mathscr{E})$ be two measurable spaces.

(1) A map $f : \Omega \to E$ is called measurable from $(\Omega, \mathscr{A})$ to $(E, \mathscr{E})$ if $f^{-1}(B) := \{\omega \in \Omega : f(\omega) \in B\} \in \mathscr{A}$ for every $B \in \mathscr{E}$, where $f^{-1}(B)$ is called the inverse image of B under f.
(2) A measurable map f from $(\Omega, \mathscr{A})$ to $(\bar{\mathbb{R}}, \bar{\mathscr{B}})$ is called a measurable function, denoted by $f \in \mathscr{A}$. A measurable map from $(\Omega, \mathscr{A})$ to $(\bar{R}^n, \bar{\mathscr{B}}^n)$ is called an n-dimensional measurable function. If f_1 and f_2 are (n-dimensional) measurable functions, then $f := f_1 + if_2$ is called an (n-dimensional) complex measurable function.

In the following, we only consider real-valued measurable functions, unless otherwise specified. Let $\mathscr{C}$ be a class of subsets of E. Then $\{f^{-1}(B) : B \in \mathscr{C}\}$ is called the inverse image of $\mathscr{C}$ under f, denoted by $f^{-1}(\mathscr{C})$ or $\sigma(f)$. Obviously, $f : (\Omega, \mathscr{A}) \to (E, \mathscr{E})$ is measurable if and only if $f^{-1}(\mathscr{E}) \subset \mathscr{A}$.

It is easy to see that the inverse is interchange with any set operations.

Property 2.3. Let f be a map from Ω to E, and let $\{B_\gamma\}_{\gamma \in \Gamma}$ be a family of subsets of Ω. Then the following can be noted:

(1) $f^{-1}(E) = \Omega,\ f^{-1}(\varnothing) = \varnothing$.
(2) $f^{-1}(B^c) = [f^{-1}(B)]^c$ for any $B \subset E$.

(3) $f^{-1}(B_1 - B_2) = f^{-1}(B_1) - f^{-1}(B_2)$ for any $B_1, B_2 \subset E$.

(4) $f^{-1}\left(\bigcup_{\gamma\in\Gamma} B_\gamma\right) = \bigcup_{\gamma\in\Gamma} f^{-1}(B_\gamma)$.

(5) $f^{-1}\left(\bigcap_{\gamma\in\Gamma} B_\gamma\right) = \bigcap_{\gamma\in\Gamma} f^{-1}(B_\gamma)$.

Property 2.4. Let $(E, \mathscr{E})$ be a measurable space. Then for any map $f : \Omega \to E$, $f^{-1}(\mathscr{E})$ is the smallest σ-algebra in Ω such that f is measurable.

Property 2.5. Let $\mathscr{C}$ be a class of subsets of E and let $f : \Omega \to E$ be a map. Then $\sigma(f) := f^{-1}(\sigma(\mathscr{C})) = \sigma(f^{-1}(\mathscr{C}))$.

Proof. Since $f^{-1}(\sigma(\mathscr{C}))$ is a σ-algebra, including $f^{-1}(\mathscr{C})$, we have $f^{-1}(\sigma(\mathscr{C})) \supset \sigma(f^{-1}(\mathscr{C}))$. So, it suffices to prove

$$\mathscr{A} := \{C \subset E : f^{-1}(C) \in \sigma(f^{-1}(\mathscr{C}))\} \supset \sigma(\mathscr{C}).$$

In fact, we have (1) $\mathscr{A} \supset \mathscr{C}$; (2) $f^{-1}(E) = \Omega \in \sigma(f^{-1}(\mathscr{C})) \Rightarrow E \in \mathscr{A}$; (3) $C \in \mathscr{A} \Rightarrow f^{-1}(C^c) = (f^{-1}(C))^c \in \sigma(f^{-1}(\mathscr{C})) \Rightarrow C^c \in \mathscr{A}$; (4) $\{C_n\}_{n\geqslant 1} \subset \mathscr{A} \Rightarrow f^{-1}\left(\bigcup_{n=1}^{\infty} C_n\right) = \bigcup_{n=1}^{\infty} f^{-1}(C_n) \in \sigma\left(f^{-1}(\mathscr{C})\right) \Rightarrow \bigcup_{n=1}^{\infty} C_n \in \mathscr{A}$. Thus, $\mathscr{A}$ is a σ-algebra including $\mathscr{C}$, hence $\mathscr{A} \supset \sigma(\mathscr{C})$. □

Theorem 2.6.

(1) *f is a real measurable function on $(\Omega, \mathscr{A})$ if and only if $\{f < x\} \in \mathscr{A}$ for every $x \in \mathbb{R}$.*
(2) *$f = (f_1, \cdots, f_n)$ is an n-dimensional function on $(\Omega, \mathscr{A})$ if and only if f_k is a real measurable function on $(\Omega, \mathscr{A})$ for $1 \leqslant k \leqslant n$.*

Proof. (1) The necessary part is obvious. To prove the sufficiency, let $\mathscr{S} = \{[-\infty, x) : x \in \mathbb{R}\}$. Then $\sigma(\mathscr{S}) = \bar{\mathscr{B}}$, so by Property 2.5, we have $f^{-1}(\bar{\mathscr{B}}) = f^{-1}(\sigma(\mathscr{S})) = \sigma(f^{-1}(\mathscr{S})) \subset \sigma(\mathscr{A}) = \mathscr{A}$. Thus, f is a measurable function on $(\Omega, \mathscr{A})$.

(2) Let f be measurable. Then for any $1 \leqslant k \leqslant n$ and $A_k \in \bar{\mathscr{B}}$, we have $\{f_k \in A_k\} = \{f \in \bar{\mathbb{R}} \times \cdots \times A_k \times \cdots \times \bar{\mathbb{R}}\} \in \mathscr{A}$. So, f_k is measurable for any $1 \leqslant k \leqslant n$. On the other hand, let f_k be measurable for any $1 \leqslant k \leqslant n$. To prove the measurability of f, we take $\mathscr{S} = \{\{f_k < r\} : 1 \leqslant k \leqslant n, r \in \mathbb{R}\}$. Since

$\bar{\mathscr{B}}^n = \sigma(\{\{x : x_k < r\} : 1 \leqslant k \leqslant n, r \in \mathbb{R}\})$, by Property 2.5, we have $f^{-1}(\bar{\mathscr{B}}^n) = \sigma(\mathscr{S})$. Combining this with the fact that the measurability of $f_k\,(1 \leqslant k \leqslant n)$ implies $\mathscr{S} \subset \mathscr{A}$, we obtain $f^{-1}(\bar{\mathscr{B}}^n) \subset \mathscr{A}$, which means that f is measurable. □

Theorem 2.7. *Let $(\Omega_i, \mathscr{A}_i), i = 1, 2, 3,$ be measurable spaces, and let $(\Omega_1, \mathscr{A}_1) \xrightarrow{f} (\Omega_2, \mathscr{A}_2) \xrightarrow{g} (\Omega_3, \mathscr{A}_3)$ be measurable maps. Then $g \circ f$ is a measurable map from $(\Omega_1, \mathscr{A}_1)$ to $(\Omega_3, \mathscr{A}_3)$.*

Proof. It follows from $(g \circ f)^{-1}(B) = f^{-1}(g^{-1}(B))$ immediately. □

By Theorem 2.6(1) and Definition 2.1, a random variable is nothing but a finite measurable function on the probability space, while Theorem 2.6(2) shows that a vector valued function is measurable if and only if each component is measurable. Theorem 2.7 says that the composition of measurable maps remains measurable.

Corollary 2.8.

(1) *Let g be a real (complex) measurable function on $(\bar{\mathbb{R}}^n, \bar{\mathscr{B}}^n)$ and $f_1, \ldots, f_n$ be real measurable functions on $(\Omega, \mathscr{A})$. Then $g(f_1, \ldots, f_n)$ is a real (complex) measurable function on $(\Omega, \mathscr{A})$.*
(2) *Let g be a real (complex) measurable function on $(\bar{\mathbb{C}}^n, \bar{\mathscr{B}_c}^n)$ and $f_1, \ldots, f_n$ be complex measurable functions on $(\Omega, \mathscr{A})$. Then $g(f_1, \ldots, f_n)$ is a real (complex) measurable function on $(\Omega, \mathscr{A})$.*

Corollary 2.9.

(1) *Let g be a real (complex) measurable function on $(\bar{\mathbb{R}}^n, \bar{\mathscr{B}}^n)$ and $f_1, \ldots, f_n$ be real random variables on $(\Omega, \mathscr{A}, \mathbb{P})$. If $\mathbb{P}(|g(f_1, \ldots, f_n)| = \infty) = 0$, then $g(f_1, \ldots, f_n)$ is a real (complex) random variable on $(\Omega, \mathscr{A}, \mathbb{P})$.*
(2) *Let g be a real (complex) measurable function on $(\bar{\mathbb{C}}^n, \bar{\mathscr{B}}_c^n)$ and $f_1, \ldots, f_n$ be complex random variables on $(\Omega, \mathscr{A}, \mathbb{P})$. If $|g(f_1, \ldots, f_n)| < \infty$, then $g(f_1, \ldots, f_n)$ is a complex random variable on $(\Omega, \mathscr{A}, \mathbb{P})$.*

2.1.2 Construction of measurable function

We first recall the measurable indicator functions with one-to-one correspondence to measurable sets and then use their combinations and limits to construct all measurable functions. This construction is fundamental for the definition of integrals, where the integral of a measurable function is regarded as the measure of the function, so that it is natural to identify the integral of an indicator function with the measure of the corresponding set.

Definition 2.10.

(1) $\forall A \subset \Omega$, its indicator function is defined by

$$\mathbf{1}_A(\omega) = \begin{cases} 1 & \text{if } \omega \in A; \\ 0 & \text{else.} \end{cases}$$

(2) Let $\{A_k\}_{1 \leqslant k \leqslant n}$ be a finite measurable partition of Ω, i.e. they are mutually disjoint sets in $\mathscr{A}$ such that $\Omega = \sum\limits_{k=1}^{n} A_k$. Then for any $a_1, \ldots, a_n \in \overline{\mathbb{R}}$, $f := \sum\limits_{k=1}^{n} a_k \mathbf{1}_{A_k}$ is called a simple function.

(3) If we take $n = \infty$ in (2) above, then f is called an elementary function.

Property 2.11.

(1) $\mathbf{1}_A$ is a measurable function on $(\Omega, \mathscr{A})$ if and only if $A \in \mathscr{A}$.

(2) Simple functions and elementary functions are all measurable functions on $(\Omega, \mathscr{A})$.

Proof. Let $f = \sum\limits_{k=1}^{n} a_k \mathbf{1}_{A_k}$ be a simple (or elementary if $n = \infty$) function. Then $\forall B \in \bar{\mathscr{B}}$ we have $f^{-1}(B) = \bigcup\limits_{k: a_k \in B} A_k \in \mathscr{A}$. □

Theorem 2.12.

(1) *A measurable function is the pointwise limit of a sequence of simple functions.*

(2) *A measurable function is the uniform limit of a sequence of elementary functions.*

(3) *A bounded measurable function is the uniform limit of a sequence of simple functions.*

(4) *A nonnegative measurable function is the (uniform) limit of a sequence of increasing simple (elementary) functions.*

Proof. (1) For $n \geqslant 1$ and $\omega \in \Omega$, let

$$f_n(\omega) = \sum_{k=-n2^n}^{n2^n-1} \frac{k}{2^n} \mathbf{1}_{\{\frac{k}{2^n} \leqslant f(\omega) < \frac{k+1}{2^n}\}} + n\mathbf{1}_{\{f(\omega) \geqslant n\}} - n\mathbf{1}_{\{f(\omega) < -n\}}.$$

Then f_n are simple functions and $|f_n - f|\mathbf{1}_{\{-n \leqslant f < n\}} < \frac{1}{2^n}$; when $f = \infty$, $f_n = n$; when $f = -\infty$, $f_n = -n$. Thus, the sequence $\{f_n\}_{n \geqslant 1}$ converges pointwise to f.

(2) For any $n \in \mathbb{N}$, let

$$f_n = \sum_{k=-\infty}^{\infty} \frac{k}{2^n} \mathbf{1}_{\{\frac{k}{2^n} \leqslant f < \frac{k+1}{2^n}\}} + \infty \mathbf{1}_{\{f=\infty\}} - \infty \mathbf{1}_{\{f=-\infty\}}.$$

Then $\{f_n\}_{n \geqslant 1}$ are elementary functions such that for any $n \geqslant 1$,

$$|f_n - f|\mathbf{1}_{\{|f|<\infty\}} < \frac{1}{2^n}; \ f_n = f \ \text{ when } |f| = \infty.$$

Thus, f_n converges uniformly to f as $n \to \infty$.

(3) If f is bounded, then by (1), the sequence $\{f_n\}_{n \geqslant 1}$ of simple functions converges uniformly to f.

(4) If f is nonnegative, then the sequences $\{f_n\}_{n \geqslant 1}$ constructed in (1) and (2) are increasing.

□

Let f be a real function on Ω. Define $f^+ = \max\{f, 0\}$ and $f^- = \max\{-f, 0\}$, which are called the positive and the negative parts of f, respectively. Then $f = f^+ - f^-, |f| = f^+ + f^-, f^+ = \frac{|f|+f}{2}$, $f^- = \frac{|f|-f}{2}$.

Theorem 2.13. *The positive part and the negative part of a measurable function are measurable. So, any measurable function can be expressed as the difference of two nonnegative measurable functions.*

2.1.3 Operations of measurable functions

Proposition 2.14. *Let $\{f_n\}_{n \geqslant 1}$ be a sequence of real functions on Ω.*

(1) *Super-limit, lower limit, supremum and infimum of $\{f_n\}_{n \geqslant 1}$ all exist, and*

$$\varliminf_{n\to\infty} f_n = \lim_{n\to\infty} \inf_{k \geqslant n} f_k = \sup_n \inf_{k \geqslant n} f_k,$$

$$\varlimsup_{n\to\infty} f_n = \lim_{n\to\infty} \sup_{k \geqslant n} f_k = \inf_n \sup_{k \geqslant n} f_k.$$

(2) *Limit of* $\{f_n\}_{n\geqslant 1}$, $\lim\limits_{n\to\infty} f_n$ *exists if and only if*

$$\forall \omega \in \Omega,\ \varliminf_{n\to\infty} f_n(\omega) = \varlimsup_{n\to\infty} f_n(\omega).$$

In this case, we denote $f_n \to f$ *as* $n \to \infty$.

A sequence of complex functions $f_n := g_n + h_n\mathrm{i}\,, n \geqslant 1$, is called convergent to $f := g + h\mathrm{i}\,$, if $g_n \to g$ and $h_n \to h$ as $n \to \infty$.

Theorem 2.15. *Let* $(\Omega, \mathscr{A})$ *be a measurable space.*

(1) *Let* $\{f_n\}_{n\geqslant 1}$ *be a sequence of real measurable functions on* $(\Omega, \mathscr{A})$. *Then*

$$\sup_{n\geqslant 1} f_n,\ \inf_{n\geqslant 1} f_n,\ \varliminf_{n\to\infty} f_n,\ \varlimsup_{n\to\infty} f_n$$

are measurable as well.

(2) *Let* $\{f_n\}_{n\geqslant 1}$ *be a sequence of complex measurable functions on* $(\Omega, \mathscr{A})$. *If* $\lim\limits_{n\to\infty} f_n$ *exists, then it is measurable.*

Proof. Note that $\forall x \in \mathbb{R}$,

$$\left\{\inf_{n\geqslant 1} f_n < x\right\} = \bigcup_{n\geqslant 1} \{f_n < x\} \in \mathscr{A}.$$

So, $\inf\limits_{n\geqslant 1} f_n$ is measurable. Since $\sup\limits_{n\geqslant 1} f_n = -\inf\limits_{n\geqslant 1}(-f_n)$, $\sup\limits_{n\geqslant 1} f_n$ is measurable.

Finally, for $x \in \mathbb{R}$,

$$\left\{\varlimsup_{n\to\infty} f_n\right\} = \bigcup_{m=1}^{\infty} \bigcup_{n=1}^{\infty} \bigcup_{k\geqslant n} \left\{f_k < x - \frac{1}{m}\right\}.$$

Thus, $\varlimsup\limits_{n\to\infty} f_n$ is measurable. As $\varliminf\limits_{n\to\infty} f_n = -\varlimsup\limits_{n\to\infty}(-f_n)$, $\varliminf\limits_{n\to\infty} f_n$ is measurable. □

Theorem 2.16. *Let* g *be a continuous function on* $D \subset \bar{\mathbb{R}}^n$. *Then* g *is a measurable function on* $(D, D \cap \bar{\mathscr{B}}^n)$. *The assertion remains true for* $\bar{\mathbb{C}}^n$ *in place of* $\bar{\mathbb{R}}^n$.

Proof. For simplicity, assume g is real. $\forall m \geqslant 1$, $\bar{R}^n$ is divided into countable many disjoint cubes with side length $1/2^m$:

$$A_{j_1,\ldots,j_n} = \left[\frac{j_1}{2^m}, \frac{j_1+1}{2^m}\right)\times\cdots\times\left[\frac{j_n}{2^m}, \frac{j_n+1}{2^m}\right), j_1,\ldots,j_n \in \mathbb{Z}\cup\{\pm\infty\},$$

For $j = -\infty$ or $+\infty$, by convention, we set $[\frac{j}{2^m}, \frac{j+1}{2^m}) = \{-\infty\}$ or $\{+\infty\}$. Rearranging these cubes, we denote them by $\{A_i^m : i, m \in \mathbb{N}\}$.

Given $x_{im} \in A_i^m$, define

$$g_m(x) = \sum_{i=1}^{\infty} \mathbf{1}_{A_i^m \cap D}(x) g(x_{im}).$$

Then g_m is measurable, and the continuity of g implies that $g_m \xrightarrow{m\to\infty} g$, so g is measurable. □

Theorem 2.17. *Let $D \subset \bar{\mathbb{C}}^n$ and $f_1, \ldots, f_n$ be measurable functions on $(\Omega, \mathscr{A})$ such that*

$$(f_1, \ldots, f_n)(\Omega) \subset D.$$

If g is a measurable function on D, then $g(f_1, \ldots, f_n)$ is measurable.

Proof. Simply note that the composition of measurable functions is measurable. □

Corollary 2.18. *The sum, difference, product and quotient of measurable functions are measurable (if the operations make sense).*

Corollary 2.19. *Let $\xi_1, \ldots, \xi_n$ be (complex) random variables on $(\Omega, \mathscr{A}, \mathbb{P})$ and let g be a finite continuous function on $\mathbb{R}^n$ ($\mathbb{C}^n$). Then $g(\xi_1, \ldots, \xi_n)$ is a (complex) random variable. Specially, the sum, difference, product and quotient of (complex) random variables are (complex) random variables (if the operations make sense).*

2.1.4 Monotone class theorem for functions

Definition 2.20. Let $\mathscr{L}$ be a family of functions on Ω such that $f \in \mathscr{L} \Rightarrow f^+, f^- \in \mathscr{L}$. A family L of functions on Ω is called an $\mathscr{L}$-system if

(1) $1 \in L$;
(2) L is closed under linear combinations;
(3) for any nonnegative and increasing sequence, $\{f_n\}_{n\geqslant 1} \subset L$, such that $f_n \uparrow f$, if either f is bounded or $f \in \mathscr{L}$, then $f \in L$.

Theorem 2.21 (Monotone class theorem for functions). *Let L be an $\mathscr{L}$-system. If L contains the indicator functions of all elements in a π-system $\mathscr{C}$, then L contains all real $\sigma(\mathscr{C})$-measurable functions in $\mathscr{L}$.*

Proof. Let $\Lambda = \{A : \mathbf{1}_A \in L\}$. Then $\Omega \in \Lambda$ and Λ is closed under the proper difference and the union of increasing sets. So, Λ is a λ-system. Since $\Lambda \supset \mathscr{C}$ and $\mathscr{C}$ is a π-system, by the monotone class theorem, we have $\Lambda \supset \sigma(\mathscr{C})$. From this and Definition 2.20(2), it follows that L contains all $\sigma(\mathscr{C})$-measurable simple functions. Let $f \in \mathscr{L}$ be $\sigma(\mathscr{C})$-measurable. Then $f^+, f^- \in \mathscr{L}$ are $\sigma(\mathscr{C})$-measurable, so that there exists a sequence of simple functions $f_n \uparrow f^+$. From this and Definition 2.20(3), it follows that $f^+ \in L$. Similarly, $f^- \in L$. Thus, $f = f^+ - f^- \in L$ by Definition 2.20(2). □

The monotone class theorem for functions is used to prove that a family F of functions have a certain property A_0. To this end, we first choose a class of functions $\mathscr{L} \supset F$ such that $L := \{f : f \text{ have property } A_0\}$ is an $\mathscr{L}$-system, then introduce a π-system $\mathscr{C}$ such that indicator functions of all subsets of $\mathscr{C}$ are contained in L, and finally verify that the family of all $\sigma(\mathscr{C})$-measurable functions includes F. Thus, by Theorem 2.21, we conclude that functions of F have property A_0.

The following theorem is an example to illustrate this procedure.

Theorem 2.22. *Let $(E, \mathscr{E})$ be a measurable space, and let $\sigma(f) = f^{-1}(\mathscr{E})$ for a map $f : \Omega \to E$. Then $\varphi : \Omega \to \bar{\mathbb{R}}$ is $\sigma(f)$-measurable if and only if there exists an $(E, \mathscr{E})$-measurable function g such that $\varphi = g \circ f$. If φ is finite (bounded), then one can take finite (bounded) g as well.*

Proof. The sufficiency follows the fact that the composition of measurable functions is measurable.

To prove the necessity, we choose $\mathscr{L}$ to be the class of all $\sigma(f)$-measurable functions on Ω, and let $L = \{g \circ f : g \in \mathscr{E}\}$. Then L is an $\mathscr{L}$-system such that the following items hold:

(1) $\mathbf{1}_\Omega = \mathbf{1}_E \circ f \in L$.
(2) $\forall g_1 \circ f, g_2 \circ f \in L$ and $a_1, a_2 \in \mathbb{R}$ such that $a_1(g_1 \circ f) + a_2(g_2 \circ f)$ makes sense, we have

$$a_1 g_1 \circ f + a_2 g_2 \circ f = [(a_1 g_1 + a_2 g_2)\mathbf{1}_A] \circ f,$$

where $A = \{x \in E : a_1 g_1(x) + a_2 g_2(x) \text{ exists}\}$. Thus, $a_1 g_1 \circ f + a_2 g_2 \circ f \in L$.
(3) If $\varphi_n \in L, \varphi_n \uparrow \varphi$, then $\forall n \geqslant 1, \exists g_n \in \mathscr{E}$ such that $\varphi_n = g_n \circ f$. Let $g = \sup\limits_{n \geqslant 1} g_n$. Then $g \in \mathscr{E}$ and $\varphi = g \circ f$, so $\varphi \in L$.

If $C \in \sigma(f)$, then there exists $B \in \mathscr{E}$ such that $C = f^{-1}(B)$, so that $\mathbf{1}_C = \mathbf{1}_B \circ f$. Thus, L contains indicator functions of all subsets of $\sigma(f)$. By Theorem 2.21, L includes $\mathscr{L}$. This proves the first assertion.

If φ is finite (bounded), and $\varphi = g \circ f$, then we can replace g by $g\mathbf{1}_{\{|g| \leqslant \|\varphi\|_\infty\}} (g\mathbf{1}_{\{|g| < \infty\}})$, so that the second assertion holds true. □

Corollary 2.23. *Let f be an n-dimensional real function on Ω. Then φ is $f^{-1}(\bar{\mathscr{B}}^n)$-measurable if and only if there exists a measurable function g on $(\bar{\mathbb{R}}^n, \bar{\mathscr{B}}^n)$ such that $\varphi = g \circ f$.*

Theorem 2.24. *Let $\mathscr{L}$ be the total of real functions on $\bar{\mathbb{R}}^n$ and L be an $\mathscr{L}$-system on $\bar{\mathbb{R}}^n$ containing all bounded continuous functions. Then L contains all Borel measurable real functions.*

Proof. Let $\mathscr{S} = \{A : A \text{ is open interval in } \bar{\mathbb{R}}^n\}$. Then $\mathscr{S}$ is a π-system and $\sigma(\mathscr{S}) = \bar{\mathbb{R}}^n$. For $A \in \mathscr{S}$, set $d(x, A^c) = \inf\{|x - y| : y \notin A\}$. $\forall m \geqslant 1$, let

$$f_m(x) = \begin{cases} 0, & x \notin A, \\ 1, & x \in A, d(x, A^c) > \frac{1}{m}, \\ md(x, A^c), & x \in A, d(x, A^c) \leqslant \frac{1}{m}. \end{cases}$$

Then f_m is continuous and $f_m \uparrow \mathbf{1}_A$, so $\mathbf{1}_A \in L$. Now, the assertion follows from Theorem 2.21. □

2.2 Distribution Function and Law

For a real function F on $\mathbb{R}^n$ and $a, b \in \mathbb{R}^n$ with $a \leqslant b$, the difference $\Delta_{b,a}F$ of F on interval $[a, b)$ is defined by $\Delta_{b,a}F := F(b) - F(a)$ when $n = 1$, and $\Delta_{b,a}F := \Delta_{b_1,a_1}\Delta_{b_2,a_2}\cdots\Delta_{b_n,a_n}$ when $n \geqslant 2$ and $a = (a_1, \ldots, a_n), b = (b_1, \ldots, b_n)$, where $\Delta_{b_i,a_i}(1 \leqslant i \leqslant n)$ is difference in the ith component.

We have the following characterization on the distribution function of a random variable.

Theorem 2.25. *Let F be an n-dimensional real function. It is the distribution function of an n-dimensional random variable if and only if the following four items hold:*

(a) *F is increasing and $\Delta_{b,a}F \geqslant 0$ for $a \leqslant b$,*
(b) *F is left-continuous,*
(c) *$F(x) \to 0$ if $\exists 1 \leqslant i \leqslant n$ such that $x_i \to -\infty$,*
(d) *$F(\infty, \infty, \ldots, \infty) := \lim_{n\to\infty} F(n, \ldots, n) = 1$.*

The necessity is obvious. So, for a function F satisfying (a)–(d), we only need to construct an n-dimensional random variable ξ on a probability space $(\Omega, \mathscr{A}, \mathbb{P})$ such that F is its distribution function. In the following, we prove a more general result for F not necessarily having properties (c) and (d). For this, we introduce a general notion of distribution functions.

Definition 2.26. A left-continuous finite real function F on $\mathbb{R}^n$ is called a distribution function if it is has nonnegative differences, i.e. $\Delta_{b,a}F \geqslant 0$ for any $a, b \in \mathbb{R}^n$ with $a \leqslant b$. In particular, F is called a probability distribution function if it satisfies (a)–(d) in Theorem 2.25.

Theorem 2.27. *Let F be a distribution function on $\mathbb{R}^n$. Then there exists a unique measure μ_F on $\mathscr{B}^n$ such that $\mu_F([a, b)) = \Delta_{b,a}F$, $a \leqslant b$. The completion of μ_F, denoted again by μ_F, is called the Lebesgue–Stieltjes (L–S) measure generated by F.*

Proof. Write $[a, b) = \prod_{k=1}^{n} [a_k, b_k)$ for $a = (a_1, \ldots, a_n) \leqslant b = (b_1, \ldots, b_n)$, where $[a_k, b_k)$ is understood as $(-\infty, b_k)$ when $a_k = -\infty$.

Let

$$\mathscr{C} = \{[a, b) : a_k \leqslant b_k, a_k \in [-\infty, +\infty), b_k \in (-\infty, +\infty], 1 \leqslant k \leqslant n\}.$$

It is clear that $\mathscr{C}$ is a semi-algebra in $\mathbb{R}^n$ and $\sigma(\mathscr{C}) = \mathscr{B}^n$.

Define a function on $\mathscr{C}$ by $\mu_F([a, b)) = \Delta_{b,a}F, a \leqslant b$. When a component of b or a is $\pm\infty$, $\mu_F([a, b))$ is understood as the limit when this component tends to $\pm\infty$, respectively. It is easy to check that μ_F is finitely additive. Since μ_F takes finite values in finite intervals, it is σ-finite. To prove that μ_F is σ-additive, let $A \in \mathscr{C}, \{A_k\}_{k\geqslant 1} \subset \mathscr{C}$ mutually disjoint and $\sum_{k=1}^{\infty} A_k = A$. It suffices to verify that $\sum_{k=1}^{\infty} \mu_F(A_k) = \mu_F(A)$.

Let μ_F be uniquely extended to a finitely additive measure on $\mathscr{F}(\mathscr{C})$, which is denoted again by μ_F. Then

$$\sum_{k=1}^{n} \mu_F(A_k) = \mu_F\left(\sum_{k=1}^{n} A_k\right) \leqslant \mu_F(A).$$

Letting $n \to \infty$, we obtain $\sum_{k=1}^{\infty} \mu_F(A_k) = \mu_F\left(\sum_{k=1}^{\infty} A_k\right) \leqslant \mu_F(A)$.

It remains to be proved that $\sum_{k=1}^{\infty} \mu_F(A_k) \geqslant \mu_F(A)$. By an approximation argument, we may assume that A is a finite interval; i.e. by first using $A^{(N)} = A \cap [-N, N)^n$ and $A_k^{(N)} = A_k \cap [-N, N)^n$ replacing A and A_k respectively for $N \in \mathbb{N}$ and then letting $N \uparrow \infty$.

Now, let $A = [a, b)$ and $A_k = [a^{(k)}, b^{(k)})$ with $a, b, a^{(k)}, b^{(k)} \in \mathbb{R}^n$ such that $a \leqslant b, a^{(k)} \leqslant b^{(k)} (k \geqslant 1)$ and $\sum_{k=1}^{\infty} A_k = A$. By the left continuity of F, $\forall \varepsilon > 0, \exists \delta > 0$ such that

$$\mu_F(A) - \varepsilon < \mu_F([a, b - \vec{\delta})),$$

where $\vec{\delta} = (\delta, \ldots, \delta)$. Moreover, for each $k \geqslant 1$, there exists $\delta^{(k)} > 0$ such that

$$\mu_F([a^{(k)} - \vec{\delta}^{(k)}, b^{(k)})) \leqslant \mu_F(A_k) + \frac{\varepsilon}{2^k}.$$

Since

$$[a, b-\vec{\delta}] \subset [a, b) = \bigcup_{k=1}^{\infty} \left(a^{(k)}, b^{(k)}\right) \subset \bigcup_{k=1}^{\infty} \left(a^{(k)} - \vec{\delta}^{(k)}, b^{(k)}\right),$$

by the finite cover theorem, we find a natural number $N \geqslant 1$ such that

$$[a, b-\vec{\delta}] \subset \bigcup_{k=1}^{N} \left(a^{(k)} - \vec{\delta}^{(k)}, b^{(k)}\right).$$

Thus,

$$\begin{aligned}\mu_F([a,b)) &\leqslant \mu_F([a, b-\vec{\delta})) + \varepsilon \leqslant \varepsilon + \sum_{k=1}^{N} \mu_F\left((a^{(k)} - \vec{\delta}^{(k)}, b^{(k)})\right) \\ &\leqslant 2\varepsilon + \sum_{k=1}^{\infty} \mu_F(A_k).\end{aligned}$$

Letting $\varepsilon \downarrow 0$, we obtain $\sum_{k=1}^{\infty} \mu_F(A_k) \geqslant \mu_F(A)$.

So far, we have proved that μ_F is a σ-finite measure on the semi-algebra $\mathscr{C}$. Then the proof is finished by the measure extension theorem (Theorem 1.42). □

It is clear that the L–S measure induced by a distribution function is finite on compact sets. Such a measure is called the Radon measure. Indeed, the inverse of Theorem 2.27 also holds, i.e. a Radon measure must be the L–S measure induced by a distribution function, see Exercise 6 at the end of this chapter.

Proof of Theorem 2.25. Let μ_F be the induced measure of F on $\mathscr{B}^n$. By (c) and (d), μ_F is a probability measure. On probability space $(\Omega, \mathscr{A}, \mathbb{P}) = (\mathbb{R}^n, \mathscr{B}^n, \mu_F)$, define the random variable $\xi(x) := x$. Then ξ is an n-dimensional random variable such that $\mathbb{P}(\xi < x) := \mu_F((-\infty, x)) = F(x)$. □

Example 2.28. Let $F(x_1, \ldots, x_n) = x_1 x_2 \ldots x_n$ for $(x_1, \ldots, x_n) \in \mathbb{R}^n$. Then F is a distribution function on $\mathbb{R}^n$ and μ_F is the Lebesgue measure on $\mathbb{R}^n$.

Definition 2.29. Let ξ be an n-dimensional random variable. The probability measure

$$(\mathbb{P} \circ \xi^{-1})(A) := \mathbb{P}(\xi \in A), \quad A \in \mathscr{B}^n$$

is called the distribution (or law, or distribution law) of ξ.

2.3 Independent Random Variables

Let T be a nonempty set. We write $S \Subset T$ if S is a nonempty finite subset of T.

Definition 2.30. Let $\{\xi^{(t)} = (\xi_{t,1}, \ldots, \xi_{t,m_t}) : t \in T\}$ be a family of random variables on $(\Omega, \mathscr{A}, \mathbb{P})$. We call $\{\xi^{(t)} : t \in T\}$ independent if for any $l \in \mathbb{N}$, $\{t_1, \ldots, t_l\} \subset T$ and $x^{(t_i)} \in \mathbb{R}^{m_{t_i}}, i = 1, \ldots, l$, there holds

$$\mathbb{P}(\xi^{(t_1)} < x^{(t_1)}, \ldots, \xi^{(t_l)} < x^{(t_l)}) = \prod_{i=1}^{l} \mathbb{P}\left(\xi^{(t_i)} < x^{(t_I)}\right).$$

The following properties are obvious.

Property 2.31.

(1) $\{\xi^{(t)} : t \in T\}$ are independent if and only if $\{\xi^{(t)} : t \in T'\}$ are independent for $\forall T' \Subset T$.
(2) Let $\bigcup\limits_{r \in I} T_r = T$ and T_r mutually disjoint with $|T_r| < \infty$. Set $\bar{\xi}^{(r)} = (\xi^{(t)} : t \in T_r)$. If $\{\xi^{(t)} : t \in T\}$ are independent, then $\{\bar{\xi}^{(r)} : r \in I\}$ are independent as well.

Due to Proposition 2.31, we only need to study the independence of finite many random variables.

Theorem 2.32. *Random variables $\{\xi^{(k)}\}_{1 \leqslant k \leqslant n}$ are independent if and only if $\forall B^{(m_k)} \in \mathscr{B}^{m_k}$,*

$$\mathbb{P}\left(\bigcap_{k=1}^{n} \{\xi^{(k)} \in B^{(m_k)}\}\right) = \prod_{k=1}^{n} \mathbb{P}\left(\xi^{(k)} \in B^{(m_k)}\right).$$

Proof. The sufficiency is obvious. By induction, we only need to prove the necessity for $n = 2$. Indeed, by Proposition 2.31(2) and the necessity for $n = 2$, we obtain, for $n = k + 1$,

$$\mathbb{P}\left(\bigcap_{i=1}^{k+1}\left\{\xi^{(i)} \in B^{(m_i)}\right\}\right) = \mathbb{P}\left(\bigcap_{i=1}^{k}\left\{\xi^{(i)} \in B^{(m_i)}\right\}\right)\mathbb{P}\left(\xi^{(k+1)} \in B^{(m_{k+1})}\right).$$

This implies the desired assertion for $n = k + 1$ by using that for $n = k$.

Now, let $n = 2$. We prove the necessity by using the monotone class theorem in two steps:

(1) Let $\mathscr{S}_k = \{(-\infty, b^k) : b^k \in \mathbb{R}^{m_k}\}$ for $k = 1, 2$. Then $\mathscr{S}_k$ is a π-system of $\mathbb{R}^{m_k}$ and $\sigma(\mathscr{S}_k) = \mathscr{B}^{m_k}$. Given $(-\infty, b) \in \mathscr{B}^{m_2}$, let

$$\mathscr{C}_1 = \left\{A_1 \in \mathscr{B}^{m_1} : \mathbb{P}\left(\xi^{(1)} \in A_1, \xi^{(2)} < b\right) = \mathbb{P}\left(\xi^{(1)} \in A_1\right)\mathbb{P}\left(\xi^{(2)} < b\right)\right\}.$$

Then $\mathscr{C}_1 \supset \mathscr{S}_1$. Now, we prove $\mathscr{C}_1$ is a λ-system.
Obvious $\Omega \in \mathscr{C}_1$. If $A^{(n)} \in \mathscr{C}_1$ and $A^{(n)} \uparrow A_1$, then $\{\xi^{(1)} \in A^{(n)}\} \uparrow \{\xi^{(1)} \in A_1\}$. By the continuity of probability, we have $A_1 \in \mathscr{C}_1$. Moreover, if $A_1 \supset A_1'$ and $A_1, A_1' \in \mathscr{C}_1$, then

$$\begin{aligned}
&\mathbb{P}\left(\xi^{(1)} \in A_1 - A_1',\ \xi^{(2)} < b\right)\\
&= \mathbb{P}\left(\xi^{(1)} \in A_1, \xi^{(2)} < b\right) - \mathbb{P}\left(\xi^{(1)} \in A_1', \xi^{(2)} < b\right)\\
&= \mathbb{P}\left(\xi^{(1)} \in A_1\right)\mathbb{P}\left(\xi^{(2)} < b\right) - \mathbb{P}\left(\xi^{(1)} \in A_1'\right)\mathbb{P}\left(\xi^{(2)} < b\right)\\
&= \mathbb{P}\left(\xi^{(1)} \in A_1 - A_1'\right)\mathbb{P}\left(\xi^{(2)} < b\right).
\end{aligned}$$

Thus, $A_1 - A_1' \in \mathscr{C}_1$, so $\mathscr{C}_1$ is a λ-system. It follows form the monotone class theorem that $\mathscr{C}_1 \supset \mathscr{B}^{m_1}$.

(2) $\forall A_1 \in \mathscr{B}^{m_1}$, let

$$\mathscr{C}_2 = \left\{A_2 \in \mathscr{B}^{m_2} : \mathbb{P}\left(\xi^{(1)} \in A_1, \xi^{(2)} \in A_2\right) = \mathbb{P}\left(\xi^{(1)} \in A_1\right)\mathbb{P}\left(\xi^{(2)} \in A_2\right)\right\}.$$

Then $\mathscr{C}_2 \supset \mathscr{S}_2$ by (1). Similar proof shows that $\mathscr{C}_2$ is a λ-system. Therefore, the proof is competed by the monotone class theorem. □

As a consequence of Theorem 2.33, the following result says that the functions of independent random variables are also independent.

Corollary 2.33. *Assume* $\{\xi^{(k)} : 1 \leqslant k \leqslant n\}$ *are independent. Let* $f_k : \mathbb{R}^{m_k} \to \mathbb{R}^{m'_k}$ *be finite Borel measurable functions. Then* $\{f_k(\xi^{(k)}) : 1 \leqslant k \leqslant n\}$ *are independent.*

Proof. $\forall A_k \in \mathscr{B}^{(m'_k)}$, we have $f_k^{-1}(A_k) \in \mathscr{B}^{(m_k)}$. Then

$$\begin{aligned}
\mathbb{P}\left(\bigcap_{k=1}^{n}\{f_k\left(\xi^{(k)}\right) \in A_k\}\right) &= \mathbb{P}\left(\bigcap_{k=1}^{n}\{\xi^{(k)} \in f_k^{-1}(A_k)\}\right) \\
&= \prod_{k=1}^{n} \mathbb{P}\left(\xi^{(k)} \in f_k^{-1}(A_k)\right) \\
&= \prod_{k=1}^{n} \mathbb{P}\left(f_k\left(\xi^{(k)}\right) \in A_k\right). \qquad \square
\end{aligned}$$

Corollary 2.34. $\{\xi^{(k)} : 1 \leqslant k \leqslant n\}$ *are independent if and only if the distribution functions of* $(\xi^{(1)}, \ldots, \xi^{(n)})$ *can be expressed as*

$$F(x^{(1)}, \ldots, x^{(n)}) = \prod_{k=1}^{n} F_k\left(x^{(k)}\right)$$

for some real function F_k *on* $\mathbb{R}^{m_k}$, *where* m_k *is the dimension of* $\xi^{(k)}, 1 \leqslant k \leqslant n$.

Proof. The necessity is obvious. To prove the sufficiency, we may assume that $\{F_k\}_{1\leqslant k\leqslant n}$ are nonnegative, otherwise simply replace F_k by $|F_k|$. Since

$$\begin{aligned}
\mathbb{P}\left(\xi^{(k)} < x^{(k)}\right) &= \mathbb{P}\left(\xi^{(k)} < x^{(k)}, \xi^{(i)} < \infty, i \neq k\right) \\
&= F\left(\infty, \ldots, x^{(k)}, \infty, \ldots, \infty\right) \\
&= F_k\left(x^{(k)}\right) \prod_{i\neq k} F_i(\infty),
\end{aligned}$$

by letting $x^{(k)} \to \infty$, we derive $\prod_{i=1}^{n} F_i(\infty) = 1$, and hence, the distribution of $\xi^{(k)}$ is given by $F_k\left(x^{(k)}\right)/F_k(\infty)$. Thus,

$$F\left(x^{(1)}, \ldots, x^{(n)}\right) = \prod_{k=1}^{n} F_k\left(x^{(k)}\right) = \prod_{k=1}^{n} \frac{F_k\left(x^{(k)}\right)}{F_k(\infty)},$$

which implies the independence of $\{\xi^{(1)}, \ldots, \xi^{(n)}\}$ by definition. □

2.4 Convergence of Measurable Functions

Let $(\Omega, \mathscr{A}, \mu)$ be a complete measure space. If some relationship holds outside a μ-null set, we say that it holds μ-almost everywhere and denote it by μ-a.e. or simply a.e. if there is no confusion. A null set is called an exception set. In this section, all measurable functions are a.e. finite.

2.4.1 Almost everywhere convergence

Definition 2.35. Let $\{f_n\}$ be a sequence of measurable functions and f be a measurable function. We say that $\{f_n\}$ converges almost everywhere to f and denote it by $f_n \xrightarrow{\text{a.e.}} f$ if there exists $N \in \mathscr{A}$ with $\mu(N) = 0$, such that $f_n(\omega) \to f(\omega), n \to \infty$ for every $\omega \notin N$.

The sequence $\{f_n\}$ is called mutually convergent almost everywhere, denoted by $f_n - f_m \xrightarrow{\text{a.e.}} 0$, if $\forall \omega \notin N$, $f_n(\omega) - f_m(\omega) \to 0$ when $n, m \to \infty$.

Obviously, $f_n - f_m \xrightarrow{\text{a.e.}} 0$ if and only if $f_{n+m} - f_n \xrightarrow{\text{a.e.}} 0$ $(n \to \infty)$ holds uniformly in $m \geqslant 1$.

Property 2.36.

(1) If $f_n \xrightarrow{\text{a.e.}} f$, then any sub-sequence $\{f_{n_k}\}$ satisfies $f_{n_k} \xrightarrow{\text{a.e.}} f$.
(2) If $f_n \xrightarrow{\text{a.e.}} f$ and $f_n \xrightarrow{\text{a.e.}} f'$, then $f = f'$ a.e.
(3) If $f_n \xrightarrow{\text{a.e.}} f$ and $g_n = f_n$ a.e., $f = g$ a.e., then $g_n \xrightarrow{\text{a.e.}} g$.
(4) If $f_n^{(k)} \xrightarrow{\text{a.e.}} f^{(k)}, k = 1, \ldots, m$, and g is a continuous function on $\bar{\mathbb{R}}^m$, then

$$g(f_n^{(1)}, \ldots, f_n^{(m)}) \xrightarrow{\text{a.e.}} g(f^{(1)}, \ldots, f^{(m)}).$$

Theorem 2.37. *Let $\{f_n\}$ be a sequence of finite measurable functions. Then there exists a finite measurable function f such that $f_n \xrightarrow{\text{a.e.}} f$ if and only if $\{f_n\}$ mutually converges almost everywhere.*

Proof. If $f_n \xrightarrow{\text{a.e.}} f$, then there exists a null set N such that $f_n(\omega) \to f(\omega), \omega \notin N$, so $\forall \omega \notin N, \{f_n(\omega)\}_{n \geqslant 1}$ is a Cauchy sequence, that is, $f_n(\omega) - f_m(\omega) \to 0$ when $n, m \to \infty$. Thus, $\{f_n\}_{n \geqslant 1}$mutually converges almost everywhere.

Conversely, if $\{f_n\}_{n \geqslant 1}$mutually converges almost everywhere, then there exists a null set N such that $\forall \omega \notin N$, $\{f_n(\omega)\}_{n \geqslant 1}$ is a Cauchy sequence, so it has a limit, denoted by $f(\omega)$. When $\omega \in N$, set $f(\omega) = 0$. Since N is measurable due to the completion of measure space, and since the limit function of measurable functions is measurable, we conclude that f is measurable and $f_n \xrightarrow{\text{a.e.}} f$. □

The following theorem follows from Definition 2.35 immediately.

Theorem 2.38. *Let $f, f_n, n \geqslant 1$ be finite measurable functions.*

(1) $f_n \xrightarrow{\text{a.e.}} f$ *if and only if* $\forall \varepsilon > 0,\ \mu\left(\bigcap_{n=1}^{\infty} \bigcup_{m=n}^{\infty} \{|f_m - f| \geqslant \varepsilon\}\right) = 0.$

In particular, when μ is finite, $f_n \xrightarrow{\text{a.e.}} f$ if and only if

$$\forall \varepsilon > 0,\ \mu\left(\bigcup_{m=n}^{\infty} \{|f_m - f| \geqslant \varepsilon\}\right) \to 0\ (n \to \infty).$$

(2) $f_n - f_m \xrightarrow{\text{a.e.}} 0$ *if and only if* $\forall \varepsilon > 0$, $\mu\left(\bigcap_{n=1}^{\infty} \bigcup_{v=1}^{\infty} \{|f_{n+v} - f_n| \geqslant \varepsilon\}\right) = 0.$

In particular, when μ is finite, $f_n - f_m \xrightarrow{\text{a.e.}} 0$ if and only if

$$\forall \varepsilon > 0,\ \mu\left(\bigcup_{v=1}^{\infty} \{|f_{n+v} - f_n| \geqslant \varepsilon\}\right) \to 0\ (n \to \infty).$$

2.4.2 Convergence in measure

Definition 2.39. A sequence $\{f_n\}_{n \geqslant 1}$ of finite measurable functions is said to converge in measure μ to a measurable function f, denoted by $f_n \xrightarrow{\mu} f$, if $\forall \varepsilon > 0, \mu(|f_n - f| \geqslant \varepsilon) \to 0(n \to \infty)$.

We call $\{f_n\}_{n\geqslant 1}$ mutually convergent in measure μ and denote $f_{n+v} - f_n \xrightarrow{\mu} 0$, if $\forall \varepsilon > 0$,

$$\sup_{v\geqslant 1} \mu(|f_{n+v} - f_n| \geqslant \varepsilon) \to 0, \quad n \to \infty,$$

Clearly, if $f_n \xrightarrow{\mu} f$, then f is finite a.e. The following properties are obvious.

Property 2.40.

(1) If $f_n \xrightarrow{\mu} f$, then any subsequence $f_{n_k} \xrightarrow{\mu} f$.
(2) If $f_n \xrightarrow{\mu} f$ and $f_n \xrightarrow{\mu} f'$, then $f = f'$ a.e.
(3) If $f_n \xrightarrow{\mu} f$ and $g_n = f_n$ a.e., $g = f$ a.e., then $g_n \xrightarrow{\mu} g$.

Theorem 2.41. *Let $f, f_n : \Omega \to \mathbb{R}^m$ be measurable and let $D \supset f(\Omega), D \supset \bigcup_{n=1}^{\infty} f_n(\Omega)$. If $g : D \to \mathbb{R}$ is uniformly continuous and $f_n \xrightarrow{\mu} f$, then $g(f_n) \xrightarrow{\mu} g(f)$.*

Proof. $\forall \varepsilon > 0, \exists \delta > 0$, we have $|g(x) - g(y)| < \varepsilon$ when $x, y \in D$ and $|x - y| < \delta$. Then $\{|g(f_n) - g(f)| \geqslant \varepsilon\} \subset \{|f_n - f| \geqslant \delta\}$. □

Corollary 2.42. *If $f_n \xrightarrow{\mu} f$ and $g_n \xrightarrow{\mu} g$, then $f_n + g_n \xrightarrow{\mu} f + g$.*

Theorem 2.43. *In Theorem 2.41, if μ is a finite measure on $(\Omega, \mathscr{A})$ and D is an open set, then g can be replaced by a continuous function.*

Proof. Let

$$D_N = \left\{ x \in \mathbb{R}^n : |x| \leqslant N, d(x, D^c) \geqslant \frac{1}{N} \right\}, \quad d(x, \varnothing) = \infty.$$

Then D_N is a bounded closed set (as $d(\cdot, D^c)$ is continuous) and $d(D_N, D_{N+1}^c) \geqslant \frac{1}{N(N+1)}$. When $N \uparrow \infty$, $D_N \uparrow D$, so that $\mu(f^{-1}(D \backslash D_N)) \downarrow 0$. Since g is uniformly continuous on D_{N+1}, $\forall \varepsilon \in (0,1)$, there exists $\delta_N > 0$ such that whenever

$\forall x, y \in D_{N+1}, |x-y| < \delta_N$, $|g(x)-g(y)| < \varepsilon$. Thus,

$$\begin{aligned}A_n := \{|g(f_n)-g(f)| \geqslant \varepsilon\} &\subset (A_n \cap \{f_n, f \in D_{N+1}\}) \\ &\quad \cup \{f \notin D_N\} \cup \left\{|f_n - f| \geqslant \frac{1}{N(N+1)}\right\} \\ &\subset \{|f_n - f| \geqslant c_N\} \cup \{f \notin D_N\},\end{aligned}$$

where $c_N := \min\{\delta_N, \frac{1}{N(N+1)}\}$. But $\lim_{n\to\infty} \mu(A_n) \leqslant 0 + \mu\left(f^{-1}(D \backslash D_N)\right)$. Let $N \uparrow \infty$ to derive

$$\lim_{n\to\infty} \mu(|g(f_n)-g(f)| \geqslant \varepsilon) = 0.$$

□

Finally, we illustrate the relationship between the a.e. convergence and the convergence in measure.

Theorem 2.44. *Let $\{f_n\}_{n\geqslant 1}$ be a sequence of finite measurable functions.*

(1) *If $f_n \xrightarrow{\mu} f$, then there exists a subsequence $\{f_{n_k}\}$ such that $f_{n_k} \xrightarrow{\text{a.e.}} f$.*
(2) *If $f_{n+v} - f_n \xrightarrow{\mu} 0$, then there exist a subsequence $\{f_{n_k}\}$ and finite measurable function f such that $f_{n_k} \xrightarrow{\text{a.e.}} f$ and $f_n \xrightarrow{\mu} f$.*
(3) *If μ is a finite measure, then $f_n \xrightarrow{\text{a.e.}} f$ implies $f_n \xrightarrow{\mu} f$.*

Proof. (1) $\forall k \geqslant 1, \exists n_k \uparrow \infty$ such that $\mu\left(|f_n - f| \geqslant 2^{-k}\right) < 2^{-k}, k \geqslant 1$. Let $f'_k = f_{n_k}$. Then $\mu\left(|f'_k - f| \geqslant 2^{-k}\right) < 2^{-k}, k \geqslant 1$. Thus, $\forall \varepsilon > 0$ and $k' \geqslant 1$ with $2^{-k'} \leqslant \varepsilon$, we have

$$\begin{aligned}\mu\left(\bigcap_{k=1}^{\infty}\bigcup_{v=1}^{\infty}\{|f'_{k+v} - f| \geqslant \varepsilon\}\right) &\leqslant \sum_{v=1}^{\infty} \mu\left(|f'_{k'+v} - f| \geqslant \varepsilon\right) \\ &\leqslant \sum_{v=1}^{\infty} 2^{-(k'+v)} = 2^{-k'}.\end{aligned}$$

Let $k' \uparrow \infty$ to derive that $f_{n_k} \xrightarrow{\text{a.e.}} f$ by Theorem 2.38(1).

(2) As in (1), we take $n_k \uparrow \infty$ such that

$$\sup_{v\geqslant 1} \mu\left(|f_{n_k+v} - f_{n_k}| \geqslant 2^{-k}\right) < 2^{-k}, \quad k \geqslant 1.$$

For $\forall \varepsilon > 0$ and $k' \geqslant 1$ with $2^{1-k'} \leqslant \varepsilon$, we have $\sum_{l=k'}^{\infty} 2^{-l} \leqslant \varepsilon$, so that $\cup_{v=1}^{\infty}\{|f'_{k'+v} - f'_{k'}| \geqslant \varepsilon\} \subset \cup_{l=k'}^{\infty}\{|f'_{l+1} - f'_l| \geqslant 2^{-l}\}$. Thus,

$$\begin{aligned}
\mu\left(\bigcap_{k=1}^{\infty}\bigcup_{v=1}^{\infty}\{|f'_{k+v} - f'_k| \geqslant \varepsilon\}\right) &\leqslant \sum_{l=k'}^{\infty} \mu\left(|f'_{l+1} - f'_l| \geqslant 2^{-l}\right) \\
&\leqslant \sum_{l=k'}^{\infty} 2^{-l} = 2^{-k'+1}.
\end{aligned}$$

By letting $k' \uparrow \infty$, it follows from Theorem 2.38(2) that f_{n_k} converges mutually almost surely, so that it converges almost surely to some finite measurable function f.

Next, we prove $f'_k \xrightarrow{\mu} f$. By $f'_k \xrightarrow{\text{a.e.}} f$, there exists a null set N such that $f'_k(\omega) \to f(\omega), \forall \omega \notin N$. Then

$$\{|f'_k - f| \geqslant \varepsilon\} \subset N \bigcup \left(\bigcup_{i=1}^{\infty}\{|f'_{k+i} - f'_{k+i-1}| \geqslant 2^{-i}\varepsilon\}\right),$$

which implies that when $\varepsilon \geqslant 2^{1-k}$,

$$\begin{aligned}
\mu(|f_k - f| \geqslant 2\varepsilon) &\leqslant \mu\left(|f'_k - f| \geqslant \varepsilon\right) + \mu(|f_{n_k} - f_k| \geqslant \varepsilon) \\
&\leqslant \sum_{i=1}^{\infty} \mu\left(|f'_{k+i} - f'_{k+i-1}| \geqslant 2^{-(k+i-1)}\right) \\
&\quad + \mu(|f_{n_k} - f_k| \geqslant \varepsilon) \\
&\leqslant 2^{1-k} + \mu(|f_{n_k} - f_k| \geqslant \varepsilon).
\end{aligned}$$

Hence, $f_n \xrightarrow{\mu} f$.

(3) Let μ be a finite measure and $f_n \xrightarrow{\text{a.e.}} f$. Then

$$\mu(|f_n - f| \geqslant \varepsilon) \leqslant \mu\left(\bigcup_{m=n}^{\infty}\{|f_m - f| \geqslant \varepsilon\}\right).$$

Combining this with $f_n \xrightarrow{\text{a.e.}} f$ and the upper continuity of measure, we obtain

$$\lim_{n\to\infty} \mu(|f_n - f| \geqslant \varepsilon) \leqslant \mu\left(\bigcap_{n=1}^{\infty}\bigcup_{m=n}^{\infty}\{|f_m - f| \geqslant \varepsilon\}\right) = 0. \qquad \square$$

Theorem 2.45. *There exists a finite measurable function f such that $f_n \xrightarrow{\mu} f$ if and only if $f_{n+v} - f_n \xrightarrow{\mu} 0$.*

Proof. The necessity follows from the triangle inequality. Following is a proof of the sufficiency.

Let $f_{n+v} - f_n \xrightarrow{\mu} 0$. By Theorem 2.44(2), there exists a subsequence such that $f_{n_k} \xrightarrow{\mu} f$ for some measurable function f. Then

$$\begin{aligned}
&\lim_{k\to\infty} \mu(|f_k - f| \geqslant \varepsilon)\\
&\quad \leqslant \lim_{k\to\infty} \mu\left(|f_k - f_{n_k}| \geqslant \frac{\varepsilon}{2}\right) + \lim_{k\to\infty} \mu\left(|f_{n_k} - f| \geqslant \frac{\varepsilon}{2}\right) = 0. \qquad \square
\end{aligned}$$

2.4.3 Convergence in distribution

Definition 2.46. Let $\{\xi_n\}_{n\geqslant 1}$ be a sequence of random variables of same dimension with the corresponding distribution functions $\{F_n\}$. Let ξ have distribution F. We call ξ_n convergent in distribution (or law) to ξ and denote $F_n \Rightarrow F$ or $\xi_n \xrightarrow{d} \xi$ if $F_n(x_0) \to F(x_0)$ holds for every continuous point x_0 of F.

Theorem 2.47. *If $\xi_n \xrightarrow{\mathbb{P}} \xi$, then $\xi_n \xrightarrow{d} \xi$.*

Proof. By $|\mathbb{P}(A) - \mathbb{P}(B)| \leqslant \mathbb{P}(A\triangle B)$, where $A\triangle B = (A - B) \cup (B - A)$ is the symmetric difference of A and B, for any unit vector $e \in \mathbb{R}^n$, we have

$$\begin{aligned}
&|F_n(x) - F(x)| = |\mathbb{P}(\xi_n < x) - \mathbb{P}(\xi < x)|\\
&= \mathbb{P}(\xi_n < x, \xi \notin (-\infty, x + \varepsilon e)) + \mathbb{P}(\xi \in (-\infty, x + \varepsilon e) \setminus (-\infty, x))\\
&\quad + \mathbb{P}(\xi_n \notin (-\infty, x), \xi < x - \varepsilon e) + \mathbb{P}(\xi \in (-\infty, x) \setminus (-\infty, x - \varepsilon e))\\
&\leqslant \mathbb{P}(|\xi - \xi_n| \geqslant \varepsilon) + \mathbb{P}(\xi \in (-\infty, x + \varepsilon e) \setminus (-\infty, x - \varepsilon e)).
\end{aligned}$$

If x is a continuous point of F, then we derive $F_n(x) \to F(x)$ by letting first $n \uparrow \infty$ and then $\varepsilon \downarrow 0$. $\square$

Corollary 2.48. *Let a be a constant. Then $\xi_n \xrightarrow{\mathbb{P}} a$ if and only if $\xi_n \xrightarrow{d} a$.*

Proof. We only need to prove the sufficiency. For simplicity, we only consider the one-dimensional case. Since the distribution for random variable $\xi \equiv a$ is $F(x) = \mathbf{1}_{(a,\infty)}$, both $a - \varepsilon$ and $a + \varepsilon$ are continuous points of F for any $\varepsilon > 0$. By $\xi_n \xrightarrow{d} a$, it follows for any $\varepsilon > 0$, $\mathbb{P}(|\xi_n - a| > \varepsilon) = \mathbb{P}(\xi_n < a - \varepsilon) + \mathbb{P}(\xi_n > a + \varepsilon) \to 0 (n \to \infty)$. □

Similarly, it is easy to check the following two assertions.

Theorem 2.49. *If $\xi_n - \xi_n' \xrightarrow{\mathbb{P}} 0$ and $\xi_n' \xrightarrow{d} \xi$, then $\xi_n \xrightarrow{d} \xi$.*

Theorem 2.50. *If $\xi_n \xrightarrow{d} \xi$ and $\eta_n \xrightarrow{d} a$, where a is a constant, then $\xi_n + \eta_n \xrightarrow{d} \xi + a$.*

2.5 Exercises

1. Prove Property 2.3.
2. Prove Corollary 2.8.
3. Prove Corollary 2.9.
4. Prove Theorem 2.13.
5. Is a distribution function increasing? Prove or disprove with a counterexample.
6. Prove that a Radon measure must be the L–S measure generated by some distribution function.
7. Prove that if $F(x) = \mathbb{P}(\xi < x)$ is continuous, then $\eta = F(\xi)$ has the uniform distribution on $[0, 1]$.
8. Prove Property 2.31.
9. Let $\{\xi_n\}_{n\geqslant 1}$ be independent with identical distribution μ. Given $A \in \mathscr{B}$ with $\mu(A) > 0$, define $\tau = \inf\{k : \xi_k \in A\}$. Prove that the distribution of ξ_τ is $\mu(\cdot \cap A)/\mu(A)$.

10. Let ξ and $\tilde{\xi}$ be independent and identically distributed. Let $\eta = \xi - \tilde{\xi}$ (which is called the symmetrization of ξ). Prove
$$\mathbb{P}(|\eta| > t) \leqslant 2\mathbb{P}(|\xi| > t/2).$$

11. Let $(\Omega, \mathscr{A}, \mathbb{P})$ be a probability space. Subclasses $\mathscr{C}_1, \ldots, \mathscr{C}_n$ of $\mathscr{A}$ are called independent if
$$\mathbb{P}\left(\bigcup_{i=1}^{n} A_i\right) = \prod_{i=1}^{n} \mathbb{P}(A_i), \quad A_i \in \mathscr{C}_i, \quad 1 \leqslant i \leqslant n.$$
Prove that if $\mathscr{C}_1, \ldots, \mathscr{C}_n$ are independent π-systems, then $\sigma(\mathscr{C}_1), \ldots, \sigma(\mathscr{C}_n)$ are independent.

12. Prove the following 0–1 laws.
    (a) Let $\{A_n\}_{n\geqslant 1}$ be a sequence of independent events and $\mathscr{T} = \bigcap\limits_{n=1}^{\infty} \sigma\{A_n, A_{n+1}, \ldots\}$. Then $P(A) = 0$ or 1 for any $A \in \mathscr{T}$.
    (b) Let $\{\xi_n\}_{n\geqslant 1}$ be a sequence of independent random variables, and let $\mathscr{T} = \bigcap\limits_{n=1}^{\infty} \sigma\{\xi_n, \xi_{n+1}, \ldots\}$, where $\sigma\{\xi_n, \xi_{n+1}, \ldots\}$ is the smallest σ-algebra in Ω such that $\{\xi_k : k \geqslant n\}$ are measurable. Then $P(A) = 0$ or 1 for any $A \in \mathscr{T}$.

13. Prove Property 2.36.

14. Prove Theorem 2.38.

15. Let $\xi_1, \xi_2, \ldots \in \{1, 2, \ldots, r\}$ be independent with identical distribution $\mathbb{P}(\xi_i = k) = p(k) > 0, 1 \leqslant k \leqslant r$. Set $\pi_n(\omega) = p(\xi_1(\omega)) \cdots p(\xi_n(\omega))$. Prove
$$-n^{-1} \log \pi_n \xrightarrow{\mathbb{P}} H := -\sum_{k=1}^{r} p(k) \log p(k),$$
where H is called Shannon's information entropy.

16. Let $\xi_n = \mathbf{1}_{A_n}$. Then $\xi_n \xrightarrow{\mathbb{P}} 0$ if and only if $\mathbb{P}(A_n) \to 0$.

17. Let $\mathscr{C}$ be a class of sets in Ω, and let f be a function on Ω. If f is $\sigma(\mathscr{C})$-measurable, then there exists a countable subclass $\mathscr{C}_f$ of $\mathscr{C}$ such that $f \in \mathscr{C}_f$.

18. If the sequence of random variables $\{\xi_n\}_{n\geqslant 1}$ is increasing and $\xi_n \xrightarrow{\mathbb{P}} \xi$, then $\xi_n \xrightarrow{\text{a.e.}} \xi$.

19. (a) If $\xi_n \xrightarrow{\text{a.e.}} \xi$, then

$$S_n := \frac{1}{n}\sum_{k=1}^{n} \xi_k \xrightarrow{\text{a.e.}} \xi.$$

(b) When $\xi_n \xrightarrow{\mathbb{P}} \xi$, does it hold that $S_n \xrightarrow{\mathbb{P}} \xi$?

20. $(\Omega, \mathscr{A}, \mathbb{P})$ is called a pure atom probability space if Ω has a partition $\{A_n\}_{n\geqslant 1}$ such that $\mathscr{A} = \sigma(\{A_n : 1 \leqslant n < \infty\})$, each $A_n (\neq \varnothing)$ is called an atom. Prove that for a sequence of random variables on a pure atom probability space, the convergence in probability is equivalent to the a.s. convergence.

21. (*Egorov's theorem*) Let $(\Omega, \mathscr{A}, \mu)$ be a finite measure space, and let f_n, f are fine measurable functions such that $f_n \xrightarrow{\text{a.e.}} f$. Then $\forall \varepsilon > 0, \exists N \in \mathscr{A}$ with $\mu(N) \leqslant \varepsilon$ such that f_n uniformly converge to f on N^c.

22. For any sequence of random variables $\{\xi_n\}$, there exists a sequence of positive numbers $\{a_n\}$ such that $a_n\xi_n \xrightarrow{\mathbb{P}} 0$.

23. Exemplify that Theorem 2.41 may fail when g is only a continuous function.

24. Prove Theorems 2.49 and 2.50.

25. Let F_n and F be distribution functions of ξ_n and ξ, respectively. If $\xi_n \xrightarrow{d} \xi$, then $\mathbb{P}(\xi_n \leqslant x) \to \mathbb{P}(\xi \leqslant x)$ and $\mathbb{P}(\xi_n > x) \to \mathbb{P}(\xi > x)$ for every continuous point x of F.

Chapter 3

Integral and Expectation

In the elementary probability theory, the expectation (also called mathematical expectation) has been defined for two typical types of random variables, i.e. by using the weighted sum with respect to the distribution sequence for a discrete random variable, and the Lebesgue integral of the product of the identity function and the distribution density function for a continuous random variable. In this chapter, we aim to define and study the expectation for general random variables on an abstract probability space $(\Omega, \mathscr{A}, \mathbb{P})$. More generally, we define the integral of a measurable function on a complete measure space $(\Omega, \mathscr{A}, \mu)$, and when $\mu = \mathbb{P}$ is a probability measure, a measurable function reduces to a random variable, and the integral is called mathematical expectation or expectation for simplicity.

Intuitively, the integral of a measurable function f with respect to μ can be regraded as the measurement result of f under μ, so the integral of a measurable indicator function $\mathbf{1}_A$ is naturally defined as $\mu(A)$. Combining this with the the construction of a measurable functions based on simple functions (Theorem 2.12), and equipping the integral with the linear property, we may define the integral for general measurable functions.

3.1 Definition and Properties for Integral

3.1.1 Definition of integral

Let $(\Omega, \mathscr{A}, \mu)$ be a complete measure space and f be a real measurable function on Ω. As explained above, for any $\mathscr{A} \in \mathscr{A}$, we call $\mu(A)$ the integral of $\mathbf{1}_A$ with respect to μ. By equipping with the linear property, we define integral for nonnegative simple functions as follows. In case an infinite function is concerned, we use the convention $0 \times \infty = 0$.

Definition 3.1. Let f be a nonnegative simple function, i.e.

$$f = \sum_{k=1}^{n} a_k \mathbf{1}_{A_k},$$

where $n \in \mathbb{N}$, $\{a_k\} \subset [0, \infty]$, and $\{A_k\} \subset \mathscr{A}$ is a partition of Ω. We call

$$\int_\Omega f \, \mathrm{d}\mu := \sum_{k=1}^{n} a_k \mu(A_k)$$

the integral of f with respect to μ. For any $A \in \mathscr{A}$, we call $\int_A f \, \mathrm{d}\mu = \int_\Omega f \mathbf{1}_A \, \mathrm{d}\mu$ the integral of f on A with respect to μ.

Clearly, the value of $\int_\Omega f \, \mathrm{d}\mu$ is independent of the expression of the simple function f and is hence well defined. As the integral is the measurement result of f under μ, we also denote $\mu(f) = \int_\Omega f \, \mathrm{d}\mu$. The following properties are obvious.

Property 3.2. Let f and g be nonnegative simple functions.

(1) (*Monotonicity*) $f \leqslant g \Rightarrow \mu(f) \leqslant \mu(g)$.
(2) (*Linearity*) $\mu(f + g) = \mu(f) + \mu(g)$ and $\mu(cf) = c\mu(f)$, $\forall c \geqslant 0$.
(3) Let $\mu_f(A) = \int_A f \, \mathrm{d}\mu$. Then μ_f is a measure on $\mathscr{A}$ and $\int_\Omega g \, \mathrm{d}(\mu \circ f^{-1}) = \int_\Omega fg \, \mathrm{d}\mu$.

By the monotonicity and the fact that a nonnegative measurable function can be approximated from below by nonnegative simple functions, we define the integrals of nonnegative measurable functions as follows.

Definition 3.3. Let f be a nonnegative measurable function. Then

$$\mu(f) = \int_\Omega f \,\mathrm{d}\mu := \sup\left\{\int_\Omega g \,\mathrm{d}\mu \colon 0 \leqslant g \leqslant f, g \text{ is a simple function}\right\}$$

is called the integral of f with respect to μ. $\forall A \in \mathscr{A}$, $\int_A f \,\mathrm{d}\mu := \mu(\mathbf{1}_A f)$ is called the integral of f on A with respect to μ.

By the definition, the monotonicity is kept by the integral of nonnegative measurable functions.

Property 3.4. If $0 \leqslant f \leqslant g$ are measurable, then $\mu(f) \leqslant \mu(g)$.

The following result is fundamental in the study of limit theorem for integrals.

Theorem 3.5 (Monotone convergence theorem). *Let* $\{f_n\}_{n\geqslant 1}$ *be nonnegative measurable functions on* Ω. *If* $f_n \uparrow f$ *as* $n \uparrow \infty$, *then*

$$\lim_{n\to\infty} \mu(f_n) = \mu(f).$$

Proof. By the monotonicity, $\mu(f_n)$ is increasing, hence its limit exists, and $f_n \leqslant f$ implies $\lim_{n\to\infty} \mu(f_n) \leqslant \mu(f)$. We only need to prove the converse inequality, i.e. for any simple function $g = \sum_{j=1}^m a_j \mathbf{1}_{A_j} + \infty \mathbf{1}_{\{f=\infty\}}$ such that $0 \leqslant g \leqslant f$, we have $\mu(g) \leqslant \lim_{n\to\infty} \mu(f_n)$. To see this, for any $\varepsilon \in (0, \min_{1\leqslant j\leqslant m} a_j)$ and $N \geqslant 1$, let

$$g_n := \sum_{j=1}^m (a_j - \varepsilon)\mathbf{1}_{A_j \cap \{|f_n - f| \leqslant \varepsilon\}} + N\mathbf{1}_{\{f=\infty, |f_n| \geqslant N\}}, \quad n \geqslant 1.$$

It is clear that $f_n \geqslant g_n$. By $f_n \to f$ and the monotonicity of integral, we obtain

$$\begin{aligned}\lim_{n\to\infty} \mu(f_n) &\geqslant \lim_{n\to\infty} \mu(g_n)\\ &= \sum_{j=1}^{m} (a_j - \varepsilon) \lim_{n\to\infty} \mu(A_j \cap \{|f_n - f| \leqslant \varepsilon\}) + N\mu(\{f = \infty, |f_n| \geqslant N\})\\ &= \sum_{j=1}^{m} (a_j - \varepsilon)\mu(A_j) + N\mu(f = \infty).\end{aligned}$$

Since ε and N are arbitrary, this implies the desired inequality $\lim_{n\to\infty} \mu(f_n) \geqslant \mu(g)$. Hence, the proof is finished. □

Finally, we define the integral of a measurable function f by the linearity and the formula $f = f^+ - f^-$, where f^+ and f^- are the positive and negative parts of f, respectively.

Definition 3.6.

(1) Let f be a measurable function on Ω. If either $\mu(f^+)$ or $\mu(f^-)$ is finite, then

$$\mu(f) = \int_\Omega f \,\mathrm{d}\mu := \mu(f^+) - \mu(f^-)$$

is called the integral of f with respect to μ. For any $A \in \mathscr{A}$ such that the integral $\mu(\mathbf{1}_A f)$ exists,

$$\int_A f \,\mathrm{d}\mu := \mu(\mathbf{1}_A f)$$

is called the integral of f on A with respect to μ. When $\mu(f)$ exists and is finite, f is called integrable (with respect to μ). To emphasize the dependence of f on x, we also denote $\mu(f) = \int_\Omega f(x)\mu(\mathrm{d}x)$ and $\mu(\mathbf{1}_A f) = \int_A f(x)\mu(\mathrm{d}x)$.

(2) Let $f = f_1 + \mathrm{i}\, f_2$ be a complex measurable function. If both $\mu(f_1)$ and $\mu(f_2)$ exist, we say that f has integral, which is defined as

$$\mu(f) = \int_\Omega f \,\mathrm{d}\mu := \int_\Omega f_1 \,\mathrm{d}\mu + \mathrm{i} \int_\Omega f_2 \,\mathrm{d}\mu.$$

we call f integrable if both f_1 and f_2 are integrable.

Proposition 3.7. *If* $f \overset{\text{a.e.}}{=} g$ *and their integrals exist, then* $\mu(f) = \mu(g)$.

3.1.2 Properties of integral

It is easy to see from Property 3.2 and Definition 3.6 that the integral has the following properties.

Theorem 3.8. *Let* f *and* g *be real measurable functions.*

(1) *Linear property*

   (a) *If the sum* $\mu(f) + \mu(g)$ *exists, then integral of* $f + g$ *exists and* $\mu(f + g) = \mu(f) + \mu(g)$.
   (b) *If* $\mu(f)$ *exists and* $A, B \in \mathscr{A}$ *are disjoint, then* $\int_{A+B} f \, \mathrm{d}\mu = \int_A f \, \mathrm{d}\mu + \int_B f \, \mathrm{d}\mu$.
   (c) *If* $c \in \mathbb{R}$ *and* $\mu(f)$ *exists, then* $\mu(cf)$ *exists, and* $\mu(cf) = c\mu(f)$.

(2) *Monotonicity*

   (a) *If* $\mu(f)$ *and* $\mu(g)$ *exist and* $f \geqslant g$, a.e., *then* $\int_A f \, \mathrm{d}\mu \geqslant \int_A g \, \mathrm{d}\mu$, $A \in \mathscr{A}$.
   (b) *If* $\mu(f)$ *exists, then* $|\mu(f)| \leqslant \mu(|f|)$.
   (c) *When* $f \geqslant 0$, $\mu(f) = 0$ *if and only if* $f = 0$, a.e.
   (d) *Let* N *be a* μ*-null set. Then* $\int_N f \, \mathrm{d}\mu = 0$.

(3) *Integrability*

   (a) f *is integrable if and only if* $\mu(|f|) < \infty$; *when* f *is integrable,* f *is finite* a.e.
   (b) *If* $|f| \leqslant g$ *and* g *is integrable, so is* f.
   (c) *If* f and g *are integrable, so is* $f + g$.
   (d) *If* $\mu(fg)$ *exists, then* $\left(\mu(fg)\right)^2 \leqslant \mu(f^2)\mu(g^2)$.

Corollary 3.9 (Markov's inequality). *If* f *is measurable and nonnegative on* $A \in \mathscr{A}$, *then* $\forall c > 0$,

$$\mu(\{f \geqslant c\} \cap A) \leqslant \frac{1}{c} \int_A f \, \mathrm{d}\mu.$$

Proof. Let $g = c\mathbf{1}_{A\cap\{f\geqslant c\}}$. Then $g \leqslant \mathbf{1}_A f$ and $\int_\Omega g\,\mathrm{d}\mu \leqslant \int_A f\,\mathrm{d}\mu$, so

$$c\mu(\{f \geqslant c\} \cap A) = \mu(g) \leqslant \int_A f\,\mathrm{d}\mu. \qquad \square$$

3.2 Convergence Theorems

As the application of the monotone convergence theorem, we have the following convergence theorems.

Theorem 3.10 (Fatou–Lebesgue theorem or Fatou's lemma). *Let g and h be real integrable functions and $\{f_n\}_{n\geqslant 1}$ be a sequence of real measurable functions.*

(1) *If $\forall n \geqslant 1, g \leqslant f_n$, then $\int_\Omega \underline{\lim}_{n\to\infty} f_n\,\mathrm{d}\mu \leqslant \underline{\lim}_{n\to\infty} \int_\Omega f_n\,\mathrm{d}\mu$.*

(2) *If $\forall n \geqslant 1, f_n \leqslant g$, then $\overline{\lim}_{n\to\infty} \int_\Omega f_n\,\mathrm{d}\mu \leqslant \int_\Omega \overline{\lim}_{n\to\infty} f_n\,\mathrm{d}\mu$.*

(3) *If $g \leqslant f_n \uparrow f$ or $\forall n \geqslant 1, g \leqslant f_n \leqslant h$,* a.e. *and $f_n \overset{\text{a.e.}}{\to} f$, then*

$$\lim_{n\to\infty} \int_\Omega f_n\,\mathrm{d}\mu = \int_\Omega f\,\mathrm{d}\mu.$$

Proof. If $g \leqslant f_n$, then $g^- \geqslant f_n^-$, so $\int_\Omega f_n^-\,\mathrm{d}\mu < \infty$, hence $\int_\Omega f_n\,\mathrm{d}\mu$ exists. A similar argument shows that $\int_\Omega f_n\,\mathrm{d}\mu$ exists in (2) and (3).

(1) Let $g_n = \inf_{k\geqslant n}(f_k - g)$. Then $g_n \geqslant 0$ and

$$g_n \uparrow \underline{\lim}_{n\to\infty}(f_n - g) = \underline{\lim}_{n\to\infty} f_n - g.$$

By the monotone convergence theorem,

$$\begin{aligned}
\int_\Omega \underline{\lim}_{n\to\infty} f_n\,\mathrm{d}\mu - \int_\Omega g\,\mathrm{d}\mu &= \lim_{n\to\infty}\int_\Omega \inf_{k\geqslant n}(f_k - g)\,\mathrm{d}\mu \\
&\leqslant \underline{\lim}_{n\to\infty}\int_\Omega (f_n - g)\,\mathrm{d}\mu = \underline{\lim}_{n\to\infty}\int_\Omega f_n\,\mathrm{d}\mu - \int_\Omega g\,\mathrm{d}\mu.
\end{aligned}$$

(2) Replacing f_n by $-f_n$ in the above proof, (2) follows from (1) immediately.

(3) When $g \leqslant f_n \uparrow f$, $0 \leqslant f_n - g \uparrow f - g$, so the assertion follows by the monotone convergence theorem. When $g \leqslant f_n \leqslant h$, a.e. and $f_n \overset{\text{a.e.}}{\to} f$, let N be a null set such that $g \leqslant f_n \leqslant h$ and $f_n \to f$ hold on N^c. Then $g\mathbf{1}_{N^c} \leqslant f_n\mathbf{1}_{N^c} \leqslant h\mathbf{1}_{N^c}$. By Theorem 3.10, (1) and (2), we have

$$\lim_{n\to\infty} \mu(f_n\mathbf{1}_{N^c}) = \mu\Big(\lim_{n\to\infty} f_n\mathbf{1}_{N^c}\Big) = \mu(f\mathbf{1}_{N^c}).$$

Combining this with Theorem 3.8, (1)(a) and (2)(d), we finish the proof. □

Theorem 3.11 (Dominated convergence theorem). *Let g be an integrable function and let $\{f_n\}$ be measurable functions such that $|f_n| \leqslant g$, a.e. for all $n \geqslant 1$. If either $f_n \xrightarrow{\text{a.e.}} f$ or $f_n \xrightarrow{\mu} f$, then $\int_\Omega f_n \,\mathrm{d}\mu \to \int_\Omega f \,\mathrm{d}\mu$.*

Proof. By Theorem 3.10(3), we only need to prove for $f_n \xrightarrow{\mu} f$. By Theorem 3.8(2)(b), it suffices to show that $\lim\limits_{n\to\infty} \int_\Omega |f_n - f| \,\mathrm{d}\mu = 0$. If this does not hold, then there exist $n_k \uparrow \infty$ and $\varepsilon > 0$ such that $\int_\Omega |f_{n_k} - f| \,\mathrm{d}\mu \geqslant \varepsilon$, $\forall k \geqslant 1$. Since $f_{n_k} \xrightarrow{\mu} f$, there exists a subsequence $f_{n'_k} \xrightarrow{\text{a.e.}} f$, so that by Theorem 3.10(3), we derive $\lim\limits_{n\to\infty} \int_\Omega |f_{n'_k} - f| \,\mathrm{d}\mu = 0$, which is a contradiction. □

Corollary 3.12. *Let $\{f_n\}_{n\geqslant 1}$ be a sequence of measurable functions. If f_n is nonnegative or $\sum\limits_{n=1}^{\infty} \int_\Omega |f_n| \,\mathrm{d}\mu < \infty$, then the integral of $\sum\limits_{n=1}^{\infty} f_n$ exists, and $\int_\Omega \sum\limits_{n=1}^{\infty} f_n \,\mathrm{d}\mu = \sum\limits_{n=1}^{\infty} \int_\Omega f_n \,\mathrm{d}\mu$.*

Proof. Let $g_n = \sum\limits_{k=1}^{n} f_k$. If f_n is nonnegative, then $g_n \uparrow \sum\limits_{n=1}^{\infty} f_n$, so the assertion follows from the monotone convergence theorem. Assume $\sum\limits_{n=1}^{\infty} \int_\Omega |f_n| \,\mathrm{d}\mu < \infty$. Let $g' = \sum\limits_{n=1}^{\infty} |f_n|$, $g'_n = \sum\limits_{k=1}^{n} |f_k|$. Then $0 \leqslant g'_n \uparrow g'$. It follows from the monotone convergence theorem

$$\sum_{n=1}^{\infty} \int_\Omega |f_n| \,\mathrm{d}\mu = \lim_{n\to\infty} \int_\Omega g'_n \,\mathrm{d}\mu = \int_\Omega g' \,\mathrm{d}\mu = \int_\Omega \sum_{n=1}^{\infty} |f_n| \,\mathrm{d}\mu,$$

so g' is integrable and $|g_n| \leqslant g'$. Since $\sum_{n=1}^{\infty} \int_\Omega |f_n| \, d\mu < \infty$ and g is a.e. finite. Hence, $g_n \xrightarrow{\text{a.e.}} \sum_{n=1}^{\infty} f_n$. Then the assertion follows from the dominated convergence theorem. □

Corollary 3.13. *If $\mu(f)$ exists, for $A \in \mathscr{A}$ and $\{A_n\}_{n=1}^{\infty} \subset \mathscr{A}$ mutually disjoint such that $A = \sum_{n=1}^{\infty} A_n$, we have $\int_A f \, d\mu = \sum_{n=1}^{\infty} \int_{A_n} f \, d\mu$.*

Proof. As $f^{\pm}\mathbf{1}_A = \sum_{n=1}^{\infty} f^{\pm}\mathbf{1}_{A_n}$, we have $\int_A f^{\pm} \, d\mu = \sum_{n=1}^{\infty} \int_{A_n} f^{\pm} \, d\mu$. Since the integral of f exists, at least one of the previous series is finite, so we can make subtraction term by term, which gives the assertion. □

The following result provides the definition of product measures.

Corollary 3.14. *Let $(\Omega_i, \mathscr{A}_i, \mu_i), 1 \leqslant i \leqslant n$, be finite many σ-finite measure spaces. Then there exists a unique σ-finite measure μ on the product measurable space $(\Omega_1 \times \cdots \times \Omega_n, \mathscr{A}_1 \times \cdots \times \mathscr{A}_n)$ such that*

$$\mu(A_1 \times \cdots \times A_n) = \mu_1(A_1) \cdots \mu_n(A_n), \quad A_i \in \mathscr{A}_i, 1 \leqslant i \leqslant n. \tag{3.1}$$

The measure μ is called the product measure of $\mu_i, 1 \leqslant i \leqslant n$, and is denoted by $\mu_1 \times \cdots \times \mu_n$.

Proof. It is easy to see that

$$\mathscr{C} := \{A_1 \times \cdots \times A_n : \ A_i \in \mathscr{A}_i, 1 \leqslant i \leqslant n\}$$

is a semi-algebra in $\Omega_1 \times \cdots \times \Omega_n$. By Theorem 1.42, it suffices to show that μ defined by (3.1) is a σ-finite measure on $\mathscr{C}$. Since each μ_i is σ-finite, so is μ. It remains to prove the σ-additivity of μ. By induction, we only prove for $n = 2$.

Let $\{A \times B, A_i \times B_i : i \geqslant 1\} \subset \mathscr{C}$ such that $A \times B = \sum_{i=1}^{\infty} A_i \times B_i$. Then for any $\omega_2 \in \Omega_2$, we have

$$\mathbf{1}_A 1_B(\omega_2) = \mathbf{1}_{A \times B}(\cdot, \omega_2) = \sum_{i=1}^{\infty} \mathbf{1}_{A_i} \mathbf{1}_{B_i}(\omega_2).$$

By Corollary 3.12 for integrals with respect to μ_1, we obtain $\mu_1(A) 1_B = \sum_{i=1}^{\infty} \mu_1(A_i) 1_{B_i}$. Applying Corollary 3.12 again for integrals with respect to μ_2, we finish the proof. □

Definition 3.15. Let f be a measurable function such that $\mu(f)$ exists. We call $\mu_f(A) := \int_A f\,\mathrm{d}\mu(A \in \mathscr{A})$ the indefinite integral of f.

It is clear that when $\mu(f^-) < \infty$, the indefinite integral μ_f is a signed measure on $\mathscr{A}$.

Proposition 3.16. *Let $(\Omega, \mathscr{A}, \mu)$ be a measure space and $\rho \geqslant 0$ be a measurable function on $(\Omega, \mathscr{A})$. If a measurable function f on $(\Omega, \mathscr{A})$ f has integral with respect to μ_ρ, then $\mu(\rho f)$ exists and $\mu(\rho f) = \mu_\rho(f)$.*

Proof. By definition, the assertion is true for f being a simple function. Combining this with the linearity of integral, Theorem 2.12(4), and Theorem 3.5, we derive the formula first for f being a simple function, then a nonnegative measurable function, and finally a measurable function such that $\mu_\rho(f)$ exists. □

Corollary 3.17. *If f is integrable, then $\int_{A_n} f\,\mathrm{d}\mu \to 0$ as $\mu(A_n) \to 0$.*

Proof. Assume $\int_{A_n} f\,\mathrm{d}\mu \nrightarrow 0$. Since $|\int_{A_n} f\,\mathrm{d}\mu| \leqslant \int |f|\,\mathrm{d}\mu < \infty$, there exists $n_k \uparrow \infty$ such that $\int_{A_{n_k}} f\,\mathrm{d}\mu \to \epsilon \neq 0$. Take a subsequence $\{n'_k\}$ of $\{n_k\}$ such that $\mu\left(A_{n'_k}\right) \leqslant \frac{1}{2^k}$. Let $B_k = \bigcup\limits_{i=k}^{\infty} A_{n'_I}$. Then $\mu(B_k) \leqslant \frac{1}{2^{k-1}}$, so $B_k \downarrow B = \bigcap\limits_{k=1}^{\infty} B_k$ is a null set. It follows that $\mathbf{1}_{A_{n'_k}} f \leqslant |\mathbf{1}_{B_k} f| \to 0$, a.e. By the dominated convergence theorem, we have $\int_{A_{n'_k}} f\,\mathrm{d}\mu \to 0$, which contradicts that $\int_{A_{n_k}} f\,\mathrm{d}\mu \to \epsilon \neq 0$ as $n_k \to \infty$. □

As applications of the dominated or monotone convergence theorem, we have the following results concerning the commutable calculations with the integral.

Corollary 3.18 (Interchange of derivative and integral). *Let $T \subset \mathbb{R}$ be an open set. If $\forall t \in T, f_t$ is integrable and $\forall \omega \in \Omega, f_t(\omega)$ is differential at t_0, then $\frac{\mathrm{d}}{\mathrm{d}t} f_t(\omega)|_{t_0}$ is measurable. If there exists an integrable function g and $\varepsilon > 0$ such that $\left|\frac{f_t - f_{t_0}}{t - t_0}\right| \leqslant g$ for $|t - t_0| < \varepsilon$, then $\left(\frac{\mathrm{d}}{\mathrm{d}t} \int_\Omega f_t\,\mathrm{d}\mu\right)|_{t_0} = \int_\Omega \frac{\mathrm{d}f_t}{\mathrm{d}t}|_{t_0}\,\mathrm{d}\mu$.*

Corollary 3.19. *Let $\{f_t\}_{t\in(a,b)}$ be a family of real integrable functions and $\frac{\mathrm{d}f_t}{\mathrm{d}t}$ exists. If there exist an integrable function g such that $\left|\frac{\mathrm{d}f_t}{\mathrm{d}t}\right| \leqslant g$, then there exists $\frac{\mathrm{d}}{\mathrm{d}t}\int f_t\,\mathrm{d}\mu = \int \frac{\mathrm{d}}{\mathrm{d}t} f_t\,\mathrm{d}\mu$ on (a,b).*

Proof. By applying the mean value theorem of differentiation, we have $\left|\frac{f_t - f_{t_0}}{t-t_0}\right| \leqslant g, t \in (a,b)$ for $\forall t_0 \in (a\,,b)$. The assertion follows from Corollary 3.18 immediately. □

Corollary 3.20 (Interchange of integrals).

(1) *Let $\{f_t\}_{t\in(a,b)}$ be a family of real integrable functions such that $\forall\omega \in \Omega$, $f_t(\omega)$ is continuous in t, and there exists an integrable function g such that $|f_t| \leqslant g$ for $\forall t \in (a,b)$. Then*

$$\int_a^b \left(\int_\Omega f_t\,\mathrm{d}\mu\right) \mathrm{d}t = \int_\Omega \left(\int_a^b f_t\,\mathrm{d}t\right) \mathrm{d}\mu.$$

(2) *If the above equation holds on finite intervals, and $\int_{-\infty}^{\infty} |f_t|\,\mathrm{d}t \leqslant h$ with h integrable, then $\int_{-\infty}^{\infty}\left(\int_\Omega f_t\,\mathrm{d}\mu\right)\mathrm{d}t = \int_\Omega \left(\int_{-\infty}^{\infty} f_t\,\mathrm{d}t\right)\mathrm{d}\mu$.*

Proof. (1) Let $a = t_0 < t_1 < \cdots < t_n = b$ be a partition of $[a,b]$. Then

$$\int_a^b f_t\,\mathrm{d}t = \lim_{n\to\infty} \sum_{i=1}^n (t_i - t_{i-1}) f_{t_i}.$$

Since $\left|\sum_{i=1}^n (t_i - t_{i-1}) f_{t_i}\right| \leqslant (b-a)g$, it follows from the dominated convergence theorem that $\int_\Omega f_t\,\mathrm{d}\mu$ is continuous in t. Thus, by the dominated convergence theorem and the linear property of integral, we have

$$\int_\Omega \left(\int_a^b f_t\,\mathrm{d}t\right) \mathrm{d}\mu = \lim_{n\to\infty} \sum_{i=1}^n (t_i - t_{i-1}) \int_\Omega f_{t_i}\,\mathrm{d}\mu = \int_a^b \left(\int_\Omega f_t\,\mathrm{d}\mu\right) \mathrm{d}t.$$

(2) Since $\int_{-\infty}^{\infty} |f_t|\,\mathrm{d}t \leqslant h$, we note that $g_n = \int_{-n}^{n} f_t\,\mathrm{d}t$ satisfies $g_n \to \int_{-\infty}^{\infty} f_t\,\mathrm{d}t$ and $|g_n| \leqslant h$. Hence, the assertion follows from the dominated convergence theorem. □

Corollary 3.21 (Interchange of summations). *Let* $\{f_{nm}\}_{n,m\geqslant 1}$ *be a family of real numbers. Assume* $f_{nm} \geqslant 0$ *or there exists a sequence of numbers* $\{g_n\}$ *such that* $\sum_{m=1}^{\infty} |f_{nm}| \leqslant g_n$ $(\forall n)$ *and* $\sum_{n=1}^{\infty} g_n < \infty$. *Then*

$$\sum_{m=1}^{\infty}\sum_{n=1}^{\infty} f_{nm} = \sum_{n=1}^{\infty}\sum_{m=1}^{\infty} f_{nm}.$$

Proof. Let $\Omega = \mathbb{N}$, μ be the counting measure on Ω and $g(n) = g_n$. Then g is integrable. Let $f_m(n) = f_{nm}$. Then $\sum_{m=1}^{\infty} |f_m| \leqslant g$. From the monotone convergence theorem or the dominated convergence theorem, it follows that

$$\sum_{m=1}^{\infty}\sum_{n=1}^{\infty} f_{nm} = \sum_{m=1}^{\infty}\int_{\Omega} f_m \,\mathrm{d}\mu = \int_{\Omega}\sum_{m=1}^{\infty} f_m \,\mathrm{d}\mu = \sum_{n=1}^{\infty}\sum_{m=1}^{\infty} f_{nm}. \qquad \square$$

Corollary 3.22. *Let* $\{f_{nm}\}_{n,m\geqslant 1}$ *be a family of real numbers such that* $0 \leqslant f_{nm} \uparrow f_n$ $(m \uparrow \infty)$ *or there exists a sequence of real numbers* $\{g_n\}_{n\geqslant 1}$ *such that* $|f_{nm}| \leqslant g_n$, $\sum_{n=1}^{\infty} g_n < \infty$, *and* $\lim_{m\to\infty} f_{nm} = f_n$. *Then*

$$\lim_{m\to\infty}\sum_{n=1}^{\infty} f_{nm} = \sum_{n=1}^{\infty} f_n.$$

3.3 Expectation

In the following, we introduce the definition of the expectation and some characters of a random variable and then establish the integral transformation formula which implies the L–S representation of expectation.

3.3.1 Numerical characters and characteristic function

Definition 3.23. Let ξ be a random variable on probability space $(\Omega, \mathscr{A}, \mathbb{P})$. If the integral of ξ with respect to $\mathbb{P}$ exists, then the integral is called the expectation of ξ, denoted by $\mathbb{E}\xi = \int_\Omega \xi \, \mathrm{d}\mathbb{P}$. If $\mathbb{E}|\xi| < \infty$, we say that ξ has finite expectation.

As in the elementary probability theory, by using expectation, we define the characteristic function and numerical characters of random variables are as follows.

Definition 3.24.

(1) Let $\xi = (\xi_1, \ldots, \xi_n)$ be an n-dimensional random variable. We call

$$\mathbb{R}^n \ni (t_1, \ldots, t_n) \mapsto \varphi_\xi(t_1, \ldots, t_n) := \mathbb{E}e^{\mathrm{i}\langle t, \xi \rangle}$$

the characteristic function of ξ, where $\mathrm{i} := \sqrt{-1}$ and $\langle t, \xi \rangle := \sum_{j=1}^n t_j \xi_j$.

(2) Let ξ be a random variable such that $\mathbb{E}\xi$ exists. Then $D\xi := \mathbb{E}|\xi - \mathbb{E}\xi|^2$ is called the variance of ξ.

(3) Let ξ be a random variable and $r > 0$. $\mathbb{E}|\xi|^r$ is called the rth moment of ξ. When $\mathbb{E}\xi$ exists, $\mathbb{E}|\xi - \mathbb{E}\xi|^r$ is called the rth central moment of ξ.

(4) Let ξ and η be two random variables such that $\mathbb{E}\xi$ and $\mathbb{E}\eta$ are finite, and

$$b_{\xi,\eta} = \mathbb{E}(\xi - \mathbb{E}\xi)(\overline{\eta - \mathbb{E}\eta})$$

exists. Then $b_{\xi,\eta}$ is called the covariance of ξ and η. If $D\xi D\eta \neq 0$ and is finite, then $r_{\xi,\eta} = \frac{b_{\xi,\eta}}{\sqrt{D\xi D\eta}}$ is called the covariance coefficient of ξ and η.

(5) Let $\xi = (\xi_1, \ldots, \xi_n)$ be an n-dimensional random variable such that $\mathbb{E}\xi = (\mathbb{E}\xi_1, \ldots, \mathbb{E}\xi_n)$ and $(b_{ij} = b_{\xi_i,\xi_j})_{1 \leqslant i,j \leqslant n}$ exist. Then

$$\mathbf{B}(\xi) = \begin{pmatrix} b_{11} & \cdots & b_{1n} \\ \vdots & & \vdots \\ b_{n1} & \cdots & b_{nn} \end{pmatrix}$$

is called the covariance matrix of ξ. If $(r_{ij} = r_{\xi_i,\xi_j})_{1 \leqslant i,j \leqslant n}$ exist, then

$$\mathbf{R}(\xi) = \begin{pmatrix} r_{11} & \cdots & r_{1n} \\ \vdots & & \vdots \\ r_{n1} & \cdots & r_{nn} \end{pmatrix}$$

is called the correlation matrix of ξ.

Besides properties listed in Theorem 3.8, the expectations for independent random variables also satisfy the following product formula.

Theorem 3.25 (Multiplication theorem). *If random variables $\xi_1, \xi_2, \ldots, \xi_n$ on probability space $(\Omega, \mathscr{A}, \mathbb{P})$ are independent, all are either nonnegative or having finite expectations, then*

$$\mathbb{E}(\xi_1\xi_2\cdots\xi_n) = \mathbb{E}\xi_1\mathbb{E}\xi_2\cdots\mathbb{E}\xi_n.$$

Proof. By induction, we only prove the formula for $n = 2$.
(1) Let ξ and η be nonnegative simple functions with $\xi = \sum_{i=1}^{n} a_i \mathbf{1}_{A_i}$ $(a_i \neq a_j, i \neq j)$ and $\eta = \sum_{i=1}^{m} b_i \mathbf{1}_{B_i}$ $(b_i \neq b_j, i \neq j)$. Then $\mathbb{P}(A_i \cap B_j) = \mathbb{P}(A_i)\mathbb{P}(B_j)$ holds for all i, j, so that

$$\xi\eta = \sum_{i=1}^{n}\sum_{j=1}^{m} a_i b_j \mathbf{1}_{A_i \cap B_j}, \quad \mathbb{E}\xi\eta = \sum_{i=1}^{n}\sum_{j=1}^{m} a_i b_j \mathbb{P}(A_i)\mathbb{P}(B_j) = \mathbb{E}\xi\mathbb{E}\eta.$$

This implies the desired formula for nonnegative ξ and η by applying Theorems 2.12(4) and 3.5.

(2) Let ξ and η have finite expectations. By Corollary 2.34, (ξ^+, ξ^-) and (η^+, η^-) are independent nonnegative random variables. Combining this with $\xi = \xi^+ - \xi^-, \eta = \eta^+ - \eta^-$, and the formula for nonnegative independent random variables proved in step (1), we finish the proof. □

The characteristic function and numerical characters have the following properties.

Proposition 3.26.

(1) *Random variables $\xi_1, \ldots, \xi_n$ are mutually independent if and only if*

$$\varphi_{(\xi_1,\ldots,\xi_n)}(t_1, \ldots, t_n) = \varphi_{\xi_1}(t_1)\cdots\varphi_{\xi_n}(t_n), \quad t_1, \ldots, t_n \in \mathbb{R}.$$

(2) *If $\xi_1, \ldots, \xi_n$ are independent variables having finite variances, then $D(\xi_1 + \cdots + \xi_n) = D\xi_1 + \cdots + D\xi_n$.*
(3) *If ξ and η are independent and having finite expectations, then $b_{\xi,\eta} = 0$.*

(4) *Let ξ be an n-dimensional random vector such that $\mathbf{B}(\xi)$ is finite. Then $\mathbf{B}(\xi) \geqslant 0$ (nonnegative definite).*
(5) *If $\mathbb{E}|\xi|^r < \infty$ for some $r > 0$, then $\mathbb{E}|\xi|^s < \infty$ holds for $s \in (0, r)$.*

Proof. We only prove (1), (4), and (5), and the rest are obvious.
(1) By Theorem 3.25, we only need to prove the sufficiency. According to Theorem 4.9 which will be proved in Chapter 4, we may construct independent random variables $\tilde{\xi}_1, \ldots, \tilde{\xi}_n$ such that their characteristic functions are $\varphi_{\xi_1}(t_1), \ldots, \varphi_{\xi_n}(t_n)$, respectively. Then $\tilde{\xi} := (\tilde{\xi}_1, \ldots, \tilde{\xi}_n)$ and $\xi := (\xi_1, \ldots, \xi_n)$ have the same characteristic function, so that by Theorem 6.4 which will be proved in Chapter 6, we know ξ and $\tilde{\xi}$ are identically distributed. From this and the mutual independence of $\tilde{\xi}_1, \ldots, \tilde{\xi}_n$, it follows that $\xi_1, \ldots, \xi_n$ are mutually independent.

(4) $\forall t_1, \ldots, t_n \in \mathbb{C}$, we have

$$\sum_{i,j=1}^{n} b_{ij} t_i \overline{t_j} = \mathbb{E}\left|\sum_{i=1}^{n} t_i(\xi_i - \mathbb{E}\xi_i)\right|^2 \geqslant 0.$$

(5) Note that $|\xi|^s \leqslant 1 + |\xi|^r$ for $0 < s < r$. □

3.3.2 Integral transformation and L–S representation of expectation

The expectation of a random variable ξ is defined as its integral with respect to the probability measure $\mathbb{P}$. Since in general the probability space $(\Omega, \mathscr{A}, \mathbb{P})$ is abstract, the expectation is not easy to calculate. Since $\mathbb{E}\xi$ is a distribution property of ξ, we aim to express it by using the integral of the identity function with respect to the distribution $\mathbb{P} \circ \xi^{-1}$ of ξ, which is a probability measure on the Euclidean space, see Definition 2.29. This is called the L–S representation of expectation.

In general, let $f : (\Omega, \mathscr{A}) \to (E, \mathscr{E})$ be a measurable map, and let μ be a measure on $(\Omega, \mathscr{A})$. Then f maps μ into the following measure $\mu \circ f^{-1}$ on $(E, \mathscr{E})$:

$$(\mu \circ f^{-1})(B) := \mu(f^{-1}(B)), \quad B \in \mathscr{E},$$

which is called the image of μ under f. We have the following integral transformation theorem.

Theorem 3.27 (Integral transformation theorem). *Let f: $(\Omega, \mathscr{A}) \longrightarrow (E, \mathscr{E})$ be measurable, let μ be a measure on $\mathscr{A}$, and let g be a measurable function on $(E, \mathscr{E})$ such that its integral with respect to $\mu \circ f^{-1}$ exists. Then the integral of $g \circ f$ with respect to μ exists and*

$$\int_{f^{-1}(B)} g \circ f \,\mathrm{d}\mu = \int_B g \,\mathrm{d}(\mu \circ f^{-1}), \quad B \in \mathscr{E}.$$

Proof. (1) Let g be an indicator function and $g = \mathbf{1}_{B'}, B' \in \mathscr{E}$. Then

$$\int_B g \,\mathrm{d}(\mu \circ f^{-1}) = (\mu \circ f^{-1})(B \cap B') = \mu(f^{-1}(B \cap B'))$$
$$= \int_{f^{-1}(B)} \mathbf{1}_{f^{-1}(B')} \,\mathrm{d}\mu = \int_{f^{-1}(B)} \mathbf{1}_{B'} \circ f \,\mathrm{d}\mu.$$

(2) By step (1), the linear property of integral, and the monotone convergence theorem, we derive the formula first for f being a simple function, then for f being a nonnegative function f, and finally for f being a measurable function such that $\mu(g \circ f)$ exists. □

In references, the L–S measure μ induced by a distribution function F is also denoted by $\mathrm{d}\mu = \mathrm{d}F$, and the associated integral is called the L–S integral.

Definition 3.28. Let μ be the Lebesgue–Stieltjes (L–S, in short) measure on $(\mathbb{R}^n, \mathscr{B}^n)$ induced by a distribution function F. Let f be a measurable function on $(\mathbb{R}^n, \mathscr{B}^n)$ such that $\mu(f)$ exists. Then the integral of f with respect to μ is called an L–S integral, denoted by

$$\mu(f) = \int_{\mathbb{R}^n} f \,\mathrm{d}\mu = \int_{\mathbb{R}^n} f \,\mathrm{d}F.$$

Let $\xi = (\xi_1, \ldots, \xi_n)$ be an n-dimensional random variable on $(\Omega, \mathscr{A}, \mathbb{P})$, having distribution function F. Then the distribution of ξ is expressed as

$$(\mathbb{P} \circ \xi^{-1})(A) := \mathbb{P}(\xi \in A) = \int_A \mathrm{d}F = \int_{\mathbb{R}^n} \mathbf{1}_A \,\mathrm{d}F, \quad A \in \mathscr{B}^n.$$

More generally, let $g := (g_1, \ldots, g_m)$ be a finite m-dimensional measurable function on $\mathbb{R}^n$. Then by Theorem 3.27, the distribution of

$\eta := g(\xi)$ is

$$(\mathbb{P}\circ\eta^{-1})(A) = \int_{g^{-1}(A)} \mathrm{d}F, \quad A \in \mathscr{B}^m.$$

This implies the following L–S representation of expectation.

Corollary 3.29. *Let $\xi = (\xi_1, \ldots, \xi_n)$ be an n-dimensional random variable on $(\Omega, \mathscr{A}, \mathbb{P})$, having distribution function F. Let $g = (g_1, \ldots, g_m)$ be a finite m-dimensional measurable function on $\mathbb{R}^n$ such that $\mathbb{E}g(\xi)$ exists. Then $\mathbb{E}g(\xi) = \int_{\mathbb{R}^n} g\,\mathrm{d}(\mathbb{P}\circ\xi^{-1}) = \int_{\mathbb{R}^n} g\,\mathrm{d}F$. In particular,*

$$\mathbb{E}\xi = \int_{\mathbb{R}^n} x\,(\mathbb{P}\circ\xi^{-1})(\mathrm{d}x) = \int_{\mathbb{R}^n} x\,\mathrm{d}F(x). \tag{3.2}$$

To conclude this section, we present the following two examples to show that the general definition of expectation covers that for discrete type and continuous type random variables presented in the elementary probability theory.

Example 3.30. Let ξ be a discrete random variable, i.e. it takes values on a countable set $\{a_i : i \geqslant 1\}$ with distribution sequence $\mathbb{P}(\xi = a_i) = p_i \geqslant 0, \sum\limits_{i=1}^{\infty} p_i = 1$. Then $(\mathbb{P}\circ\xi^{-1})(A) = \sum\limits_{a_i\in A} p_i$ holds for any $A \in \mathscr{B}$. By (3.2) and noting that the identity function satisfies

$$x = \sum_{i=1}^{\infty} a_i \mathbf{1}_{\{a_i\}}(x), \quad (\mathbb{P}\circ\xi^{-1})\text{-a.s.},$$

we obtain

$$\mathbb{E}\xi = \int_{\mathbb{R}} x\,(\mathbb{P}\circ\xi^{-1})(\mathrm{d}x) = \sum_{i=1}^{\infty} a_i p_i.$$

Example 3.31. Let ξ be a continuous type random variable with distribution density function ρ such that $\mathbb{E}\xi$ exists. Then its distribution is the indefinite integral of ρ with respect to the Lebesgue measure $\mathrm{d}x$, so that by Propositions 3.16 and (3.2), we obtain $\mathbb{E}\xi = \int_{\mathbb{R}} x\,(\mathbb{P}\circ\xi^{-1})(\mathrm{d}x) = \int_{\mathbb{R}} x\rho(x)\,\mathrm{d}x$.

3.4 L^r-space

Definition 3.32. Let $(\Omega, \mathscr{A}, \mu)$ be a measure space and let $r \in (0, \infty)$.

$$L^r(\mu) := \{f : f \text{ is a measurable function on } \Omega,\ \mu(|f|^r) < \infty\}$$

is called the L^r-space of μ. A sequence $\{f_n\} \subset L^r(\mu)$ is said to converge in $L^r(\mu)$ to some measurable function f if $\mu(|f_n - f|^r) \to 0$ $(n \to \infty)$, which is denoted by $f_n \xrightarrow{L^r(\mu)} f$.

To ensure the uniqueness of limit, an element in $L^r(\mu)$ is regarded as an equivalent class in the sense of μ-a.e. equal, that is, we identify two functions f and g in $L^r(\mu)$ if $f = g$ μ-a.e.

Let $\|f\|_r = \mu(|f|^r)^{1/(r \wedge 1)}$. We will prove when $r \in (0,1)$, $(L^r(\mu), \|\cdot\|_r)$ is a complete metric space with distance $d_r(f,g) := \|f - g\|_r$; when $r \geqslant 1$, it is a Banach space. In particular, $L^2(\mu)$ is a Hilbert space with inner product $\langle f, g\rangle := \mu(fg)$. To prove this assertion, we first recall some classical inequalities, then extend them to integrals of functions, and finally compare the convergence in L^r with the convergences in a.e. and in measure. Moreover, the space $(L^r(\mu), \|\cdot\|_r)$ is separable if $\mathscr{A}$ is generated by an at most countable sub-class of $\mathscr{A}$, which we will not prove in this textbook.

3.4.1 Some classical inequalities

Proposition 3.33. *If $a \geqslant 0, b \geqslant 0, 0 < \alpha < 1, \alpha + \beta = 1$, then $a^\alpha b^\beta \leqslant a\alpha + b\beta$ and the equality holds if and only if $a = b$.*

Proof. As log is a concave function, $\log(a\alpha + b\beta) \geqslant \alpha \log a + \beta \log b = \log(a^\alpha b^\beta)$, and clearly the equality holds if and only if $a = b$. □

Proposition 3.34 (Hölder's inequality). *Let $r > 1, \frac{1}{r} + \frac{1}{s} = 1$. Then*

$$\mu(|f\,g|) \leqslant (\mu(|f|^r))^{\frac{1}{r}} (\mu(|g|^s))^{\frac{1}{s}}. \tag{3.3}$$

The equality holds if and only if $\exists c_1, c_2 \in \mathbb{R}$ with $|c_1| + |c_2| > 0$ such that

$$c_1|f|^r + c_2|g|^s = 0, \ \mu\text{-a.e.} \tag{3.4}$$

Proof. The inequality is obvious when $f = 0$ or $g = 0$ or the R.H.S. is infinite, the inequality holds obviously. So, we may and do assume $0 < \mu(|f|^r), \mu(|g|^s) < \infty$. In this case, let

$$a = \frac{|f|^r}{\mu(|f|^r)}, \quad b = \frac{|g|^s}{\mu(|g|^s)}, \quad \alpha = \frac{1}{r}, \quad \beta = \frac{1}{s},$$

From Proposition 3.33, it follows that

$$\frac{|f\,g|}{\|f\|_r\|g\|_s} \leqslant \frac{|f|^r}{r\mu(|f|^r)} + \frac{|g|^s}{s\mu(|g|^s)},$$

where the equality holds if and only if $\frac{|f|^r}{\mu(|f|^r)} = \frac{|g|^s}{\mu(|g|^s)}$. Combining this with Theorem 3.8, we may take integrals with respect to μ in both sides to derive (3.3), and the equality holds if and only if $\frac{|f|^r}{\mu(|f|^r)} = \frac{|g|^s}{\mu(|g|^s)}$ holds μ-a.e., which implies (3.4) for $c_1 = \frac{1}{\mu(|f|^r)}$ and $c_2 = -\frac{1}{\mu(|g|^s)}$. Finally, it is clear that (3.4) implies the equality in (3.3). □

Corollary 3.35 (Jensen's inequality). $\forall r > 1, \mathbb{E}|\xi| \leqslant (\mathbb{E}|\xi|^r)^{\frac{1}{r}}$, *and the equality hold if and only if* $|\xi|^r$ *is* a.s. *constant.*

Proposition 3.36 (C_r-inequality).

$$\forall a_1, \ldots, a_n \in \mathbb{R}, \ |a_1 + \cdots + a_n|^r \leqslant n^{(r-1)^+}(|a_1|^r + \cdots + |a_n|^r).$$

When $r > 1$, *the equality holds if and only if* $a_1 = \cdots = a_n$; *when* $r = 1$, *the equality holds if and only if* a_i *have same signs; when* $r < 1$, *the equality holds if and only if at most one of* $\{a_i\}$ *is not zero.*

Proof. (1) Case $r > 1$. Let $\Omega = \{1, \ldots, n\}$, $\mathscr{A} = 2^\Omega$ equipped with probability measure $\mathbb{P}(A) = \frac{1}{n}|A|$, where $|A|$ is the number of points in A. Consider the random variable $\xi(i) := a_i, 1 \leqslant i \leqslant n$. Then $\mathbb{E}|\xi| = \frac{1}{n}\sum_{i=1}^{n}|a_i|, \mathbb{E}|\xi|^r = \frac{1}{n}\sum_{i=1}^{n}|a_i|^r$. By Jensen's inequality, $n^{-r}\left(\sum_{i=1}^{n}|a_i|\right)^r \leqslant \frac{1}{n}\sum_{i=1}^{n}|a_i|^r$, hence the inequality holds, and the equality holds if and only if $|\xi|$ are constant, i.e. $|a_i| = |a_j|, \forall i, j$. But $\left|\sum_{i=1}^{n} a_i\right| = \sum_{i=1}^{n} |a_i|$ if and only if a_i have same signs, so $a_i = a_j, \forall i, j$.

(2) Case $r \leqslant 1$. We only prove a_i are not all null. Note

$$\frac{|a_k|}{\sum_{i=1}^{n} |a_i|} \leqslant \frac{|a_k|^r}{\left(\sum_{i=1}^{n} |a_i|\right)^r}, \quad r \leqslant 1.$$

Make summation in k on both sides to derive the inequality. When $r = 1$, the equality holds if and only if a_i have same signs. And when $r \leqslant 1$, the equality if and only if $\forall k, |a_k| / \sum_{i=1}^{n} |a_i| = 1$ or 0, i.e. only one of a_i is not null. □

Proposition 3.37 (C_r-inequality). *Let $f_1, \ldots, f_n$ be measurable functions. Then*

$$\mu(|f_1 + \cdots + f_n|^r) \leqslant n^{(r-1)^+} \sum_{i=1}^{n} \mu(|f_i|^r),$$

and the equality hold if and only if

(1) *when $r > 1$, $\forall i \neq j, f_i = f_j$, a.e.;*
(2) *when $r < 1$, at most one of $\mu(|f_i|)$ is not null;*
(3) *when $r = 1$, f_i, a.e. have the same sign.*

Proposition 3.38 (Minkowski's inequality). *Let $r \geqslant 1, f, g \in L^r(\mu)$. Then*

$$(\mu|f + g|^r)^{\frac{1}{r}} \leqslant (\mu|f|^r)^{\frac{1}{r}} + (\mu|g|^r)^{\frac{1}{r}},$$

and the equality holds if and only if

(1) *when $r > 1$, there exist $c_1, c_2 \in \mathbb{R}$ not all null and having the same signs such that $c_1 f - c_2 g = 0$, a.e.;*
(2) *when $r = 1$, f, g have the same sign, a.e.*

Proof. Since the assertion for $r = 1$ is trivial, we only consider $r > 1$. By Hölder's inequality, we obtain

$$\mu(|f+g|^r) \leqslant \mu(|f||f+g|^{r-1}) + \mu(|g||f+g|^{r-1})$$
$$\leqslant \|f\|_r(\mu(|f+g|^r))^{\frac{r-1}{r}} + \|g\|_r(\mu(|f+g|^r))^{\frac{r-1}{r}},$$

and the equality holds if and only if there exist $c_1, c_2, c_3, c_4 \in \mathbb{R}$ with $|c_1|+|c_2|>0$ and $|c_3|+|c_4|>0$ such that μ-a.e $|f|^r c_1 + |f+g|^r c_2 = c_3|g|^r + c_4|f+g|^r = 0$ and f, g have the same sign. This implies the desired assertion. □

3.4.2 Topology property of $L^r(\mu)$

Theorem 3.39. *Let $r \in (0,\infty)$. Then $L^r(\mu)$ is a complete metric space under the distance $d_r(f,g) := \|f-g\|_r$. Moreover, it is a Banach space when $r \geqslant 1$ and a Hilbert space when $r = 2$.*

Proof. (a) Clearly, $\|f\|_r = 0$ if and only if $f = 0, \mu$-a.e. Thus, $\forall f \in L^r(\mu), \|f\|_r = 0$ if and only if $f = 0$. Obviously, $L^r(\mu)$ is obviously a linear space and d_r satisfies the triangle inequality by C_r-inequality (for $r < 1$) or Minkowski's inequality (for $r \geqslant 1$). Moreover, when $r \geqslant 1, \|\cdot\|_r$ is a norm.

(b) It remains to prove the completeness. Let $\{f_n\}$ be a Cauchy sequence in $(L^r(\mu), d_r)$. Then $\forall \varepsilon > 0$, by Khinchin's inequality, we derive that when $n, m \to \infty$,

$$\mu(|f_n - f_m| \geqslant \varepsilon) \leqslant \frac{1}{\varepsilon^r}\mu(|f_n - f_m|^r) \to 0.$$

Thus, $\{f_n\}$ converges mutually in measure. By Theorem 2.45, there exists a subsequence $n_k \uparrow \infty$ such that $f_{n_k} \xrightarrow{\text{a.e.}} f(\text{say})$. It follows that for $\forall m \geqslant 1$, $f_m - f_{n_k} \xrightarrow{\text{a.e.}} f_m - f$ $(n_k \to \infty)$. Then Fatou's lemma (Theorem 3.10) implies

$$\mu(|f_m - f|^r) = \mu\left(\varliminf_{n_k\to\infty} |f_m - f_{n_k}|^r\right) \leqslant \varliminf_{n_k\to\infty} \mu\left(|f_m - f_{n_k}|^r\right).$$

Since $\{f_n\}$ is a Cauchy sequence in $L^r(\mu)$, by letting $m \to \infty$, we obtain $\lim_{m\to\infty} \mu(|f_m - f|^r) = 0$. □

Proposition 3.40.

(1) *Let μ be a finite measure. If* $f_n \xrightarrow{L^r(\mu)} f$, *then* $f_n \xrightarrow{L^{r'}(\mu)} f, r' \in (0, r)$.
(2) *If* $f_n \xrightarrow{L^r(\mu)} f$, *then* $\mu(|f_n|^r) \to \mu(|f|^r)$.

Proof. (1) and (2) follow from Hölder's inequality and the triangle inequality of d_r, respectively. □

3.4.3 Links of different convergences

Definition 3.41. Let $(\Omega, \mathscr{A}, \mu)$ be a finite measure space, and let $\{f_t, t \in T\}$ be a family of real measurable functions on Ω.

(1) $\{f_t, t \in T\}$ is called uniformly continuous in integral if

$$\lim_{\mu(A)\to 0} \sup_{t\in T} \mu(|f_t|\mathbf{1}_A) = 0.$$

(2) $\{f_t, t \in T\}$ is called uniformly integrable if

$$\lim_{n\to\infty} \sup_{t\in T} \mu(|f_t|\mathbf{1}_{\{|f_t|\geqslant n\}}) = 0.$$

Theorem 3.42. *Let μ be a finite measure and $\{f_n\}_{n\geqslant 1} \subset L^r(\mu)$. The following statements are equivalent:*

(1) $f_n \xrightarrow{L^r(\mu)} f$.
(2) $f_n \xrightarrow{\mu} f$ *and* $\{|f_n - f|^r\}_{n\geqslant 1}$ *is uniformly continuous in integral.*
(3) $f_n \xrightarrow{\mu} f$ *and* $\{|f_n|^r\}_{n\geqslant 1}$ *is uniformly continuous in integral.*
(4) $f_n \xrightarrow{\mu} f$ *and* $\{|f_n|^r\}_{n\geqslant 1}$ *is uniformly integrable.*

Proof. (a) First, we prove (1) ⇔ (2).
(1) ⇒ (2) Since $\mu(|f_n - f| \geqslant \varepsilon) \leqslant \varepsilon^{-r}\mu(|f_n - f|^r)$, (1) implies $f_n \xrightarrow{\mu} f$. To prove that $\{|f_n - f|^r\}_{n\geqslant 1}$ is uniformly continuous in integral, for $\forall \varepsilon > 0$, take $n_\varepsilon \geqslant 1$ such that $\mu(|f_n - f|^r) < \varepsilon$ for $\forall n \geqslant n_\varepsilon$. Then we

have

$$\sup_{n\geqslant 1}\int_A |f_n - f|^r \,\mathrm{d}\mu \leqslant \varepsilon + \sum_{n=1}^{n_\varepsilon} \mu(\mathbf{1}_A|f_n - f|^r).$$

As for given n, $\lim\limits_{\mu(A)\to 0} \mu(\mathbf{1}_A|f_n - f|^r) = 0$, so

$$\overline{\lim_{\mu(A)\to 0}} \sup_{n\geqslant 1} \mu(\mathbf{1}_A|f_n - f|^r) \leqslant \varepsilon.$$

Since ε is arbitrary, $\{|f_n - f|^r\}_{n\geqslant 1}$ is uniformly continuous in integral.
(2) $\Rightarrow$ (1) Let $A_n = \{|f_n - f| \geqslant \varepsilon\}$. Then $\mu(A_n) \to 0$ and the uniform continuity in integral implies $\mu(\mathbf{1}_{A_n}|f_n - f|^r) \leqslant \sup\limits_{m\geqslant 1} \mu(\mathbf{1}_{A_n}|f_m - f|^r) \to 0, n \to \infty$. Hence,

$$\overline{\lim_{n\to\infty}} \mu(|f_n - f|^r) \leqslant \overline{\lim_{n\to\infty}} \mu(|f_n - f|^r \mathbf{1}_{\{|f_n - f|\geqslant\varepsilon\}}) + \varepsilon^r \mu(\Omega) = \varepsilon^r \mu(\Omega).$$

Since ε is arbitrary, we have $f_n \xrightarrow{L^r(\mu)} f$.

(b) Again, by Theorem 2.44, $f_n \xrightarrow{\mu} f$ implies that there exists a subsequence $\{f_{n_k}\}$ such that $f_{n_k} \xrightarrow{\text{a.e.}} f$. Thus, by Fatou's lemma,

$$\forall A \in \mathscr{A}, \mu(f\mathbf{1}_A) \leqslant \underline{\lim}_{k\to\infty} \mu(f_{n_k}\mathbf{1}_A) \leqslant \sup_n \mu(f_n\mathbf{1}_A).$$

Since $|\|\mathbf{1}_A f_n\|_r - \|\mathbf{1}_A(f_n - f)\|_r| \leqslant \|\mathbf{1}_A f\|_r$, the uniform continuity in integral of $\{|f_n - f|^r\}_{n\geqslant 1}$ is equivalent to that of $\{|f_n|^r\}_{n\geqslant 1}$. That is, (2) $\Leftrightarrow$ (3).

(c) From the equivalence of (2) and (3), it follows that $\{\mu(|f_n|)\}_{n\geqslant 1}$ is bounded, so that the uniformly continuity of integrals of $\{|f_n|^r\}_{n\geqslant 1}$ is equivalent to the uniform integrability of $\{|f_n|^r\}_{n\geqslant 1}$(cf. Exercise 25 in this chapter). Hence, (3) $\Leftrightarrow$ (4). □

Noting that $|\mu(f_n) - \mu(f)| \leqslant \mu(|f_n - f|)$ and $f_n \xrightarrow{a.e.} f$ implies $f_n \xrightarrow{\mu} f$ for finite μ, we have the following consequence of Theorem 3.42.

Corollary 3.43 (Dominated convergence). *Let $(\Omega, \mathscr{A}, \mu)$ be a finite measure space, and let $\{f_n, f : n \geqslant 1\}$ be uniformly integrable measurable functions. If $f_n \xrightarrow{\mu} f$, then $\mu(f_n) \to \mu(f)$.*

3.5 Decompositions of Signed Measure

Basing on the decomposition $f = f^+ - f^-$ for a function, we aim to formulate a signed measure φ as the differences of two measures φ^+ and φ^-. This is called Hahn's decomposition, from which we will introduce Lebesgue's decomposition which uniquely expresses a signed measure as the sum of an indefinite integral part and a singular part. The uniqueness of Lebesgue's decomposition leads to the Radon–Nikodym derivative between measures, which is crucial to develop the analysis on the space of measures. By applying Lebesgue's decomposition to L–S measure, we decompose a distribution function into the discrete part, the absolutely continuous part, and the singular part, which classifies random variables into three types: the discrete type, the continuous type and the singular type, where the first two types have been studied in the elementary probability theory.

3.5.1 Hahn's decomposition theorem

To decompose a signed measure as the difference of two measures, we consider the indefinite integral μ_f for a measurable function f with $\mu(f^-) < \infty$, for which the natural decomposition is $\mu_f = \mu_{f^+} - \mu_{f^-}$. To define φ^+ and φ^- for a general signed measure φ, we reformulate μ_{f^+} and μ_{f^-} as follows:

$$\mu_{f^-}(A) = \mu_f(A \cap D), \quad \mu_{f^+}(A) = \mu_f(A \cap D^c), \quad A \in \mathscr{A}, \ D := \{f \leqslant 0\}.$$

It is clear that $\mu_f(D) = \inf_{A \in \mathscr{A}} \mu_f(A)$. This indicates that for a signed measure φ, if we could find a set $D \in \mathscr{A}$ reaching $\inf_{A \in \mathscr{A}} \varphi(A)$, we could define $\varphi^+(A)$ and $\varphi^-(A)$ as $\varphi(A \cap D^c)$ and $\varphi(A \cap D)$, respectively. So, we first prove the existence of D.

Theorem 3.44. *Let φ be a signed measure on $(\Omega, \mathscr{A})$. Then $\exists D \in \mathscr{A}$ such that*

$$\varphi(D) = \inf_{A \in \mathscr{A}} \varphi(A).$$

Proof. Take $\{A_n\}$ such that $\varphi(A_n) \downarrow \inf_{A \in \mathscr{A}} \varphi(A)$. Since $\inf_{A \in \mathscr{A}} \varphi(A) \leqslant 0$, we may assume that $\varphi(A_n)$ are finite. Let $A = \bigcup_{n=1}^{\infty} A_n$.

For any $k \geqslant 1$, we have $A = A_k + (A - A_k) =: A_{k,1} + A_{k,2}$. $\forall n \geqslant 2$, we can write

$$A = \sum_{i_1,i_2,\ldots,i_n=1}^{2} A_{1,i_1} \cap A_{2,i_2} \cap \cdots \cap A_{n-1,i_{n-1}} \cap A_{n,i_n}.$$

As n is increasing, the partitions become finer and finer. For each partition, we take out the subsets with negative φ-values. Intuitively, the union of such subsets approaches the desired set D when $n \to \infty$. To confirm this observation, for each $n \geqslant 1$, let

$$B_n = \sum_{\substack{1 \leqslant i_1,i_2,\ldots,i_n \leqslant 2 \\ \varphi(A_{1,i_1} \cap A_{2,i_2} \cap \cdots \cap A_{n,i_n}) \leqslant 0}} A_{1,i_1} \cap A_{2,i_2} \cap \cdots \cap A_{n,i_n} =: \sum_{i=1}^{k_n} A'_{n,i}.$$

By the σ-additivity of φ, we have $\varphi(B_n) \leqslant \varphi(A_n)$. Let

$$D = \bigcap_{n=1}^{\infty} \bigcup_{k=n}^{\infty} B_k = \lim_{n\to\infty} \bigcup_{k=n}^{\infty} B_k.$$

As $(n+1)$th partition is finer than that of nth, a subset $A'_{n+1,i}$ of B_{n+1} is either included by B_n or disjoint with B_n. Then for any $m > n$, we have

$$B_n \cup \cdots \cup B_m = B_n + \sum_{A'_{n+1,i} \cap B_n = \varnothing} A'_{n+1,i} + \sum_{A'_{n+2,i} \cap (B_n \cup B_{n+1}) = \varnothing} A'_{n+2,i} + \cdots + \sum_{A'_{m,i} \cap (B_n \cup \cdots \cup B_{m-1}) = \varnothing} A'_{m,i}.$$

Thus, by the σ-additivity of φ and $\varphi(A'_{i,j}) \leqslant 0$, it follows $\varphi(B_n \cup \cdots \cup B_m) \leqslant \varphi(B_n) \leqslant \varphi(A_n)$. Letting $m \uparrow \infty$, we obtain $-\infty < \varphi\left(\bigcup_{k=n}^{\infty} B_k\right) \leqslant \varphi(A_n)$ by the lower continuity of signed measures (note $\varphi(A_n)$ is finite). Finally, the upper continuity of signed measure implies that

$$\varphi(D) = \lim_{n\to\infty} \varphi\left(\bigcup_{k=n}^{\infty} B_k\right) \leqslant \lim_{n\to\infty} \varphi(A_n) = \inf_{A \in \mathscr{A}} \varphi(A). \qquad \square$$

Corollary 3.45. *Let φ be a signed measure on a measurable space $(\Omega, \mathscr{A})$. Then there exists $D \in \mathscr{A}$ such that $\varphi(A \cap D) = \inf_{B \in A \cap \mathscr{A}} \varphi(B)$ and $\varphi(A \cap D^c) = \sup_{B \in A \cap \mathscr{A}} \varphi(B)$ for any $A \in \mathscr{A}$.*

Proof. Let $D \in \mathscr{A}$ such that $\varphi(D) = \inf_{A \in \mathscr{A}} \varphi(A), \varphi(D^c) = \sup_{A \in \mathscr{A}} \varphi(A)$. Then $\forall A \in \mathscr{A}$ and $B \in A \cap \mathscr{A}$, we have $\varphi(A \cap D) + \varphi(D - A) = \varphi(D) \leqslant \varphi(B \cup (D - A)) = \varphi(B) + \varphi(D - A)$. Since both $\varphi(D) \leqslant 0$, $\varphi(A \cap D)$ and $\varphi(D - A)$ are finite, $\varphi(A \cap D) \leqslant \varphi(B)$. Thus, $\inf_{B \in A \cap \mathscr{A}} \varphi(B) \leqslant \varphi(A \cap D) \leqslant \inf_{B \in A \cap \mathscr{A}} \varphi(B)$, i.e. $\varphi(A \cap D) = \inf_{B \in A \cap \mathscr{A}} \varphi(B)$.

On the other hand, $\forall B \in A \cap \mathscr{A}, \varphi(A \cap D^c) + \varphi(A \cap D) = \varphi(A) = \varphi(B) + \varphi(B^c \cap A)$. Since $\varphi(A \cap D) = \inf_{B \in A \cap \mathscr{A}} \varphi(B)$ is finite, $\varphi(A \cap B^c) \geqslant \inf_{B \in A \cap \mathscr{A}} \varphi(B) = \varphi(A \cap D)$, so $\varphi(A \cap D^c) = \varphi(B) + \varphi(B^c \cap A) - \varphi(A \cap D) \geqslant \varphi(B)$. Hence, $\sup_{B \in A \cap \mathscr{A}} \varphi(B) \leqslant \varphi(A \cap D^c) \leqslant \sup_{B \in A \cap \mathscr{A}} \varphi(B)$, i.e. $\varphi(A \cap D^c) = \sup_{B \in A \cap \mathscr{A}} \varphi(B)$. □

By Corollary 3.45, we have the following theorem.

Theorem 3.46 (Hahn's decomposition theorem). *Let φ be a σ-additive function on $(\Omega, \mathscr{A})$, and let*

$$\varphi^+(A) = \sup_{\mathscr{A} \ni B \subset A} \varphi(B), \quad \varphi^-(A) = -\inf_{\mathscr{A} \ni B \subset A} \varphi(B), \quad A \in \mathscr{A}.$$

Then both φ^+ and φ^- are measures on $\mathscr{A}$, and $\varphi = \varphi^+ - \varphi^-$.

The formula $\varphi = \varphi^+ - \varphi^-$ is called Hahn's decomposition of φ, where φ^+ and φ^- are called the positive and negative parts of φ, respectively. Moreover, the measure $|\varphi| := \varphi^+ + \varphi^-$ is called the total variation measure of φ. Note that in general $|\varphi(A)| \neq |\varphi|(A)$ for $A \in \mathscr{A}$.

By Corollary 3.45 and Theorem 3.46, if $\varphi = \mu_f$ for some measurable function f with $\mu(f^-) < \infty$, then $\varphi^+ = \mu_{f^+}$ and $\varphi^- = \mu_{f^-}$, as suggested in the beginning of this part.

3.5.2 Lebesgue's decomposition theorem

Let $(\Omega, \mathscr{A}, \mu)$ be a measure space. If $\varphi = \mu_f$ is the indefinite integral of a measurable function f with $\mu(f^-) < \infty$, then $|\varphi|(N) = 0$ holds for any μ-null set N. We will prove the converse result, i.e. a signed measure having this property must be the indefinite integral of a measurable function with respect to μ. To this end, we introduce the following notions and establish Lebesgue's decomposition theorem.

Definition 3.47. Let $(\Omega, \mathscr{A}, \mu)$ be a measure space, and let φ be a signed measure on $\mathscr{A}$.

(1) φ is called absolutely continuous with respect to μ, denoted by $\varphi \ll \mu$, if $\varphi(N) = 0$ holds for all μ-null set $N \in \mathscr{A}$.
(2) We call φ and μ (mutually) singular if there exists $N \in \mathscr{A}$ such that $\mu(N) = 0$ and $|\varphi|(N^c) = 0$.

Theorem 3.48. *Let $(\Omega, \mathscr{A}, \mu)$ be a σ-finite measure space, and let φ be a σ-finite signed measure. Then $\varphi \ll \mu$ if and only if there exists a measurable function f such that $\mu(f^-) < \infty$ and $\varphi = \mu_f$.*

The sufficient part is obvious, and the necessary part is implied by the following Lebesgue's decomposition theorem.

Theorem 3.49 (Lebesgue's decomposition theorem). *Let μ and φ be as in Theorem 3.48. Then φ is uniquely decomposed as $\varphi = \varphi_c + \varphi_s$, where φ_c is the indefinite integral of a measurable function with respect to μ and φ_s is a signed measure singular to μ. The composition is unique.*

Proof. (1) *Uniqueness of decomposition*

Consider two such decompositions: $\varphi = \varphi_c + \varphi_s = \varphi_c' + \varphi_s'$. Then there are μ-null sets N_1, N_2 such that $|\varphi_s|(N_1^c) = |\varphi_s'|(N_2^c) = 0$. Let $N = N_1 \cup N_2$. We have $\mu(N) = 0$ and $|\varphi_s|(N^c) = |\varphi_s'|(N^c) = 0$. So, $\forall A \in \mathscr{A}$,

$$\varphi_s(A \cap N) = (\varphi_c + \varphi_s)(A \cap N) = (\varphi_c' + \varphi_s')(A \cap N) = \varphi_s'(A \cap N),$$

and $\varphi_s(A \cap N^c) = \varphi_s'(A \cap N^c) = 0$. Thus, $\varphi_s = \varphi_s'$. Similarly, we have $\varphi_c(A \cap N) = \varphi_c'(A \cap N) = 0$ and

$$\varphi_c(A \cap N^c) = (\varphi_c + \varphi_s)(A \cap N^c) = (\varphi_c' + \varphi_s')(A \cap N^c) = \varphi_c'(A \cap N^c),$$

so that $\varphi_c = \varphi_c'$. Then the decomposition is unique.

(2) *Existence of decomposition*
(i) Assume that μ and φ are finite measures. Let

$$\Phi = \left\{ f : f \geqslant 0, \int_A f \,\mathrm{d}\mu \leqslant \varphi(A), \forall A \in \mathscr{A} \right\}, \quad \alpha = \sup_{f \in \Phi} \mu(f).$$

Clearly, Φ is not empty and $\alpha \in [0, \varphi(\Omega)]$. Take $\{f_n\}_{n\geqslant 1} \subset \Phi$ such that $\alpha_n := \mu(f_n) \uparrow \alpha \leqslant \varphi(\Omega) < \infty$. Set $g_n = \sup\limits_{k\leqslant n} f_k$. Then $0 \leqslant g_n \uparrow f := \sup\limits_{k\geqslant 1} f_k$. For given $n \geqslant 1$, put $A_k = \{\omega : g_n(\omega) = f_k(\omega)\}$ $(1 \leqslant k \leqslant n)$. Then $\bigcup\limits_{k=1}^{n} A_k = \Omega$. Moreover, let $B_k = A_k - \bigcup\limits_{i=1}^{k-1} A_I$. Then $\{B_k\}$ are mutually disjoint and $\bigcup\limits_{k=1}^{n} B_k = \Omega$. Thus, $\forall A \in \mathscr{A}$,

$$\int_A g_n \,\mathrm{d}\mu = \sum_{k=1}^{n} \int_{A\cap B_k} f_k \,\mathrm{d}\mu \leqslant \sum_{k=1}^{n} \varphi(B_k \cap A) = \varphi(A).$$

Hence, $\int_A f \,\mathrm{d}\mu \leqslant \varphi(A)$. By this and definition of α, it follows that $\mu(f) = \alpha$.

Let

$$\varphi_c(A) = \int_A f \,\mathrm{d}\mu, \ \varphi_s(A) = \varphi(A) - \int_A f \,\mathrm{d}\mu.$$

For any $n \geqslant 1$, let $\varphi_n = \varphi_s - \frac{\mu}{n}$. From the proof of Hahn's decomposition theorem, it follows that there exists $D_n \in \mathscr{A}$ such that

$$\varphi_n(D_n \cap A) \leqslant 0, \ \varphi_n(D_n^c \cap A) \geqslant 0, \ \forall A \in \mathscr{A}.$$

Let $D = \bigcap\limits_{n=1}^{\infty} D_n$. Then $\forall n$,

$$D \subset D_n, \varphi_s(D \cap A) \leqslant \frac{1}{n}\mu(D \cap A).$$

Thus, $\varphi_s(D \cap A) = 0$ for any $A \in \mathscr{A}$.

To prove $\mu(D^c) = 0$, it suffices to show $\mu(D_n^c) = 0$ for any n. In fact,

$$\begin{aligned}\int_A \left(f + \frac{1}{n}\mathbf{1}_{D_n^c}\right) \mathrm{d}\mu &= \varphi_c(A) + \frac{1}{n}\mu(A \cap D_n^c)\\ &= \varphi(A) - \varphi_s(A) + \frac{1}{n}\mu(A \cap D_n^c)\\ &= \varphi(A) - \varphi_n(A \cap D_n^c) - \varphi_s(A \cap D_n)\\ &\leqslant \varphi(A) - \varphi_s(A \cap D_n) \leqslant \varphi(A).\end{aligned}$$

It follows that $f+\frac{1}{n}\mathbf{1}_{D_n^c} \in \Phi$. Thus, $\alpha \geqslant \int_\Omega (f+\frac{1}{n}\mathbf{1}_{D_n^c})\,\mathrm{d}\mu = \int_\Omega f\,\mathrm{d}\mu + \frac{1}{n}\mu(D_n^c) = \alpha$. This together with $\mu(f) = \alpha$ implies $\frac{1}{n}\mu(D_n^c) = 0$, hence $\mu(D_n^c) = 0$.

(ii) Let μ and φ be σ-finite measures. There exists $\{A_n\}_{n\geqslant 1}$ mutually disjoint such that $\bigcup\limits_{n=1}^{\infty} A_n = \Omega$ and $\mu(A_n), \varphi(A_n) < \infty\,(\forall n)$. From (i), it follows that there exists $\varphi_c^{(n)}$ and $\varphi_s^{(n)}$ such that

$$\begin{aligned}\varphi(A_n \cap \bullet) &= \varphi_c^{(n)}(A_n \cap \bullet) + \varphi_s^{(n)}(A_n \cap \bullet),\\ \varphi_c^{(n)}(A_n \cap \bullet) &= \int_{A_n \cap \bullet} f^{(n)} d\mu.\end{aligned}$$

Let N_n be a μ-null set such that $\varphi_s^{(n)}(N_n^c \cap A \cap A_n) = 0$ for any $A \in \mathscr{A}$. Set

$$f = \sum_{n=1}^{\infty} \mathbf{1}_{A_n} f^{(n)},\ \varphi_c(A) = \int_A f\,\mathrm{d}\mu,\ \varphi_s(A) = \sum_{n=1}^{\infty} \varphi_s^{(n)}(A_n \cap A).$$

Again, let $N = \bigcup\limits_{n=1}^{\infty} N_n$. Then $\forall A \in \mathscr{A}$, and we have $\varphi_s^{(n)}(N^c \cap A \cap A_n) \leqslant \varphi_s^{(n)}(N_n^c \cap A \cap A_n) = 0$. It follows that $\varphi_s(N^c \cap A) = \sum\limits_n {\varphi_s}^{(n)}(N^c \cap A \cap A_n) = 0$. Hence, φ_s and μ are singular.

(3) *General case.* From Hahn's decomposition theorem, we have $\varphi = \varphi^+ - \varphi^-$. But by (ii), we have φ^+ and φ^- have decompositions $\varphi^+ = {\varphi_c}^+ + {\varphi_s}^+, \varphi^- = {\varphi_c}^- + {\varphi_s}^-$. Then $\varphi = ({\varphi_c}^+ - {\varphi_c}^-) + ({\varphi_s}^+ - {\varphi_s}^-)$ □

As a direct consequence of Lebesgue's decomposition theorem, we have the following result.

Theorem 3.50 (Radon–Nikodym theorem). *Let* $(\Omega, \mathscr{A}, \mu)$ *be a* σ*-finite measure space, and let* φ *be a* σ*-finite signed measure on* $\mathscr{A}$*. If* φ *is absolutely continuous with respect to* μ*, then there exists a* μ-a.e. *unique measurable function* f *such that* $\varphi = \mu_f$.

This result can be extended to not σ-finite signed measures.

Theorem 3.51 (*Generalization of the Radon–Nikodym theorem*). *Let* $(\Omega, \mathscr{A}, \mu)$ *be a* σ*-finite measure space and* φ *be a signed measure on* $\mathscr{A}$*. If* φ *is absolutely continuous with respect to* μ*, then there exists a* μ-a.e. *unique measurable function* f *such that* $\varphi = \mu_f$.

Proof. By the σ-finiteness of μ and Hahn's decomposition of φ, we may assume that μ is a finite measure and φ is a measure. In this case, we apply Theorem 3.50 with Ω replaced by the largest measurable set on which the restriction of φ is σ-finite, then the function f is defined as ∞ outside this set.

Let

$$\mathscr{C} = \{A \in \mathscr{A} : \varphi \text{ is } \sigma\text{-finite on } A\}.$$

Set $s = \sup\limits_{B \in \mathscr{C}} \mu(B)$ and then take $\{B_n\}_{n \geqslant 1} \subset \mathscr{C}$ such that $\mu(B_n) \uparrow s$. Let $B = \bigcup\limits_{n=1}^{\infty} B_n$. Then $B \in \mathscr{C}$ and $s = \mu(B)$. Since φ is σ-finite on $B \cap \mathscr{A}$, Theorem 3.50 implies that there exists f_1 such that $\varphi(A \cap B) = \int_{A \cap B} f_1 \, \mathrm{d}\mu, A \in \mathscr{A}$. Let

$$f(\omega) = \begin{cases} f_1(\omega), & \omega \in B, \\ \infty, & \omega \notin B. \end{cases}$$

Then for any $A \in \mathscr{A}$ with $\mu(A \cap B^c) > 0$, we have $\int_A f \, \mathrm{d}\mu = \infty$. On the other hand, if $\mu(A \cap B^c) > 0$, then $\varphi(A \cap B^c) = \infty$. If not, then φ is σ-finite on $B \cup A$ and $\mu(B \cup A) = \mu(B^c \cap A) + \mu(B) > s$, which contradicts the fact $s = \sup\limits_{B \in \mathscr{B}} \mu(B)$. Thus, $\forall A \in \mathscr{A}$,

$$\int_A f \, \mathrm{d}\mu = \int_{A \cap B} f \, \mathrm{d}\mu + \int_{A \cap B^c} f \, \mathrm{d}\mu = \varphi(A \cap B) + \infty \cdot \mu(A \cap B^c) = \varphi(A).$$

□

The above theorem leads to the notion of Radon–Nikodym derivative.

Definition 3.52. Let $(\Omega, \mathscr{A}, \mu)$ be a σ-finite measure space and let φ be a signed measure which is absolutely continuous with respect to μ. Then there exists a μ-a.e. unique measurable function f such that $\varphi = \mu_f$ (equivalently, $\mathrm{d}\varphi = f\,\mathrm{d}\mu$). The function f is called the Radon–Nikodym derivative of φ with respect to μ and is denoted by $\frac{\mathrm{d}\varphi}{\mathrm{d}\mu}$.

The following result is reformulated from Proposition 3.16.

Corollary 3.53. *Let ν and μ be σ-finite measures on $\mathscr{A}$ such that $\nu \ll \mu$. If f is a measurable function, then the integral of f with respect to ν exists if and only if the integral of $f\frac{\mathrm{d}\nu}{\mathrm{d}\mu}$ with respect to μ exists, and $\int_A f\,\mathrm{d}\nu = \int_A f\frac{\mathrm{d}\nu}{\mathrm{d}\mu}\,\mathrm{d}\mu$ for any $A \in \mathscr{A}$.*

3.5.3 Decomposition theorem of distribution function

By applying Lebesgue's decomposition theorem to the L–S measure induced by a distribution function with respect to the Lebesgue measure $\mathrm{d}x$, we derive the following decomposition theorem for distribution function.

Theorem 3.54. *Any distribution function F on $\mathbb{R}^n$ can be decomposed as the sum of three distribution functions: $F = F_c + F_d + F_s$, where the L–S measure induced by F_c is absolutely continuous with respect to $\mathrm{d}x$, the L–S measure induced by F_d is supported on a finite or countable set, and the L–S measure induced by F_s is singular with $\mathrm{d}x$ and having null measure on singletons. This decomposition is unique in the sense of induced L–S measures. The functions F_c, F_d and F_s are called the absolute part, the discrete part and the singular part of F, respectively.*

Proof. Let μ be the L–S measure induced by F. By the Lebesgue's decomposition theorem, we have $\mu = \mu_c + \mu_s'$, where $\mu_c \ll \mathrm{d}x, \mu_s'$ is singular with $\mathrm{d}x$. Let $A = \{x \in \mathbb{R}^n : \mu_s'(\{x\}) > 0\}$. Then A is at most countable. Define $\mu_d(B) = \sum_{x \in B\cap A} \mu_s'\{x\}$ and let F_d be distribution function of μ_d. Finally, let $\mu_s = \mu_s' - \mu_d$. Then μ_s is singular with $\mathrm{d}x$, and $\mu_s(\{x\}) = 0$ for every $x \in \mathbb{R}^n$, which has distribution function $F_s = F - F_c - F_d$.

Uniqueness of decomposition follows from that of Lebesgue decomposition and properties of F_s and F_d. □

3.6 Exercises

1. For a nonnegative measurable function f on a measure space $(\Omega, \mathscr{A}, \mu)$, let
$$\bar{\int_\Omega} f \, \mathrm{d}\mu = \inf\left\{\int_\Omega g \, \mathrm{d}\mu : g \geqslant f, g \text{ is a simple function}\right\}.$$
Exemplify that $\bar{\int_\Omega} f \, \mathrm{d}\mu$ and $\int_\Omega f \, \mathrm{d}\mu$ may not be identical.

2. Prove Theorem 3.8.

3. Exemplify that when f is a complex measurable function whose integral exists and c is a complex, the integral of cf may not exist. What happens when f is integrable?

4. Let f be a complex measurable function such that $\mu(f)$ exists. Prove $|\int_\Omega f \, \mathrm{d}\mu| \leqslant \int_\Omega |f| \, \mathrm{d}\mu$.

5. Exemplify that we cannot get rid of dominated condition $g \leqslant f_n$ in Theorem 3.10(1).

6. Let $\{f_{nm}\}_{n,m \geqslant 1}$ be a family of nonnegative numbers. Prove
$$\varliminf_{m\to\infty} \sum_{n=1}^{\infty} f_{nm} \geqslant \sum_{n=1}^{\infty} \varliminf_{m\to\infty} f_{nm}.$$

7. Exemplify that for a sequence of random variables, the convergence in $L^r(\mathbb{P})$ does not imply the a.s. convergence and vice verse.

8. Let $(\Omega, \mathscr{A}, \mu)$ be a measure space, and let φ be a finite signed measure such that $\varphi \ll \mu$. Then $\varphi(A_n) \to 0$ for any $\{A_n\} \subset \mathscr{A}$ with $\mu(A_n) \to 0$. Exemplify that this assertion does not hold when φ is σ-finite.

(*Hint: Let* $\Omega = (0,1), \mu = \mathrm{d}x, \varphi = \mu_f, f(x)\frac{1}{x}, A_n = (0, \frac{1}{n})$.)

9. If $\{\xi_n\}$ converges in distribution to ξ, then $\mathbb{E}|\xi| \leqslant \varliminf_{n\to\infty} E|\xi_n|$.

10. Prove Corollary 3.35.

11. Prove Proposition 3.37.

12. Let $\xi \geqslant 0$ such that $\mathbb{E}\xi^2 < \infty$. Prove $\mathbb{P}(\xi > 0) \geqslant \frac{(\mathbb{E}\xi)^2}{\mathbb{E}\xi^2}$.

13. Let $A_1, \ldots, A_n$ be events and $A = \bigcup\limits_{i=1}^{n} A_i$. Prove

    (a) $\mathbf{1}_A \leqslant \sum\limits_{i=1}^{n} \mathbf{1}_{A_i}$.

    (b) $\mathbb{P}(A) \geqslant \sum\limits_{i=1}^{n} \mathbb{P}(A_i) - \sum\limits_{i<j} \mathbb{P}(A_i \cap A_j)$.

    (c) $\mathbb{P}(A) \leqslant \sum\limits_{i=1}^{n} \mathbb{P}(A_i) - \sum\limits_{i<j} \mathbb{P}(A_i \cap A_j) + \sum\limits_{i<j<k} \mathbb{P}(A_i \cap A_j \cap A_k)$.

14. Apply Jensen's inequality to prove (geometric mean is dominated by algebraic mean): $a_1, \ldots, a_n \geqslant 0$ and $\alpha_1, \ldots, \alpha_n \geqslant 0$ such that $\alpha_1 + \cdots + \alpha_n = 1$, we have $\prod\limits_{i=1}^{n} a_i^{\alpha_i} \leqslant \sum\limits_{i=1}^{n} \alpha_i a_i$.

15. Let $\xi > 0$. Prove

$$\lim_{t\to\infty} t \int_{[\xi>t]} \frac{1}{\xi} \,\mathrm{d}\mathbb{P} = 0, \quad \lim_{t\to 0} t \int_{[\xi>t]} \frac{1}{\xi} \,\mathrm{d}\mathbb{P} = 0.$$

16. Let ξ and η be independent random variables with distribution functions F and G, respectively. Formulate the distribution function of $\xi + \eta$ using F and G.

17. Random variables ξ and η are independent if and only if for any f, g,

$$\mathbb{E}f(\xi)g(\eta) = \mathbb{E}f(\xi)\mathbb{E}g(\eta).$$

18. (a) If events $A_1, A_2, \ldots$ satisfy $\sum\limits_{n=1}^{\infty} \mathbb{P}(A_n) < \infty$, then $\mathbb{P}(\bigcap\limits_{n=1}^{\infty} \bigcup\limits_{k=n}^{\infty} A_k) = 0$.

    (b) If events $A_1, A_2, \ldots$ are independent and $\sum\limits_{n=1}^{\infty} \mathbb{P}(A_n) = \infty$, then $\mathbb{P}(\bigcap\limits_{n=1}^{\infty} \bigcup\limits_{k=n}^{\infty} A_k) = 1$.

19. Let $p_n \in [0, 1)$. Apply the previous exercise to prove $\prod\limits_{n=1}^{\infty} (1 - p_n) = 0$ if and only if $\sum\limits_{n=1}^{\infty} p_n = \infty$.

20. Let ξ be a random variable taking values of nonnegative integers. Prove $\mathbb{E}\xi = \sum\limits_{n \geqslant 1} \mathbb{P}(\xi \geqslant n)$. What happens when ξ takes values of integers?

21. Let ξ be a nonnegative random variable. Prove $\mathbb{E}\xi = \int_0^{\infty} \mathbb{P}(\xi \geqslant x)\, \mathrm{d}x$. What happens for a general real random variable?

22. Prove $\xi_n \xrightarrow{\mathbb{P}} \xi$ if and only if

$$\mathbb{E}\left(\frac{|\xi_n - \xi|}{1 + |\xi_n - \xi|}\right) \to 0.$$

23. Let $\phi \geqslant 0$ such that $\lim\limits_{x \to \infty} \phi(x)/x \to \infty$ and T be an index set. If $\mathbb{E}\phi(|\xi_t|) \leqslant C < \infty$ for any $t \in T$, then $\{\xi_t, t \in T\}$ is uniformly integrable.

24. Let $\{f_n\}$ be a sequence of real measurable functions on $(\Omega, \mathscr{A}, \mu)$. If $\sup\limits_{n \geqslant 1} \mu(|f_n|^r) < \infty$ for some $r > 0$, then $\forall s \in (0, r)$, $\{|f_n|^s\}$ is uniformly continuous in integral.

25. Let $(\Omega, \mathscr{A}, \mu)$ be a finite measure space, and let $\{f_t : t \in T\}$ be a family of random variables such that $\{\mu(|f_t|) : t \in T\}$ is bounded. Then $\{f_t\}_{t \in T}$ is uniformly continuous in integral if and only if it is uniformly integrable.

26. Let $r \in (0, \infty)$ and let $(\Omega, \mathscr{A}, \mu)$ be a measure space. Prove that the class of integrable simple functions is dense in $L^r(\mu)$.

27. Let $1/p + 1/q = 1, p, q > 1$. Prove $\|f\|_p = \{\mu(fg) : \|g\|_q \leqslant 1\}$.

28. If a sequence of random variables $\{\xi_n\}_{n\geqslant 1}$ is uniformly bounded, then ξ_n converges in probability if and only if it converges in $L^r(\mathbb{P})$, where $r \in (0, \infty)$.

29. For a measurable function f, define the **essential supremum** by

$$\|f\|_\infty = \inf\{M : \mu(\{\omega : |f(\omega)| > M\}) = 0\}.$$

   (a) Prove that $\|\cdot\|_\infty$ satisfies the triangle inequality.
   (b) If $\mu(\Omega) < \infty$, then $\|f\|_\infty = \lim_{r\to\infty} \|f\|_r$.

30. Assume $\mathbb{E}\xi^2 < \infty$. Prove that $\mathbb{E}\xi$ attains the minimum of $\mathbb{E}(\xi - c)^2$ over $c \in \mathbb{R}$.

31. If ξ and η are independent random variables having finite expectations, $\mathbb{E}|\xi + \eta|^2 < \infty$. Prove $\mathbb{E}(|\xi|^2 + |\eta|^2) < \infty$.

32. Let ξ be a random variable with $m := \mathbb{E}\xi \in \mathbb{R}$ and $\sigma^2 = D\xi \in (0, \infty)$.
   (a) Prove

$$\mathbb{P}(\xi - m \geqslant t) \leqslant \frac{\sigma^2}{\sigma^2 + t^2}, \quad t \geqslant 0.$$

   (b) Prove

$$\mathbb{P}(|\xi - m| \geqslant t) \leqslant \frac{2\sigma^2}{\sigma^2 + t^2}, \quad t \geqslant 0.$$

33. Let f be a convex function on $\mathbb{R}$, and let ξ be a random variable with finite expectation. Prove that $\mathbb{E}f(\xi)$ exists and $f(\mathbb{E}\xi) \leqslant \mathbb{E}f(\xi)$.

34. Let ξ be a random variable with finite expectation. If there is a strictly convex function ϕ on $\mathbb{R}$ such that $\mathbb{E}\phi(\xi) = \phi(\mathbb{E}\xi)$, then ξ is an a.s. constant.

35. If independent random variables ξ and η have distributions μ and ν, respectively, then for $A \in \mathscr{B}$ and $B \in \mathscr{B}^2$,

$$\mathbb{P}((\xi, \eta) \in B) = \int_{\mathbb{R}} \mathbb{P}((x, \eta) \in B)\mu(\mathrm{d}x) = \int_{\mathbb{R}} \mathbb{P}((\xi, y) \in B)\nu(\mathrm{d}y)$$

and

$$\mathbb{P}(\xi \in A, (\xi, \eta) \in B) = \int_A \mathbb{P}((x, \eta) \in B)\mu(\mathrm{d}x).$$

36. (*Uniform distribution on Cantor sets*) Let C be the Cantor set defined in Exercise 31 of Chapter 2. We call

$$F(x) := \begin{cases} 0, & x \leqslant 0, \\ 1, & x \geqslant 1, \\ \frac{1}{2}, & x \in [\frac{1}{3}, \frac{2}{3}], \\ \frac{1}{4}, & x \in [\frac{1}{9}, \frac{2}{9}], \\ \frac{3}{4}, & x \in [\frac{7}{9}, \frac{8}{9}], \\ \cdots & \cdots \end{cases}$$

the uniform distribution function on C. Prove

(a) F is continuous;
(b) F is singular with Lebesgue measure.

37. Let μ_1 and μ_2 be finite signed measures. Set $\mu_1 \vee \mu_2 = \mu_1 + (\mu_2 - \mu_1)^+, \mu_1 \wedge \mu_2 = \mu_1 - (\mu_1 - \mu_2)^+$. Then $\mu_1 \vee \mu_2$ is the minimal signed measure such that $\nu \geqslant \mu_i (i = 1, 2)$ and $\mu_1 \wedge \mu_2$ is the maximal signed measure such that $\nu \leqslant \mu_i (i = 1, 2)$.

38. Let μ be a σ-finite measure on $(\Omega, \mathscr{A})$ such that $\mathscr{A}$ contains all singletons. Then the set

$$\{x \in \Omega : \mu(\{x\}) > 0\}$$

is at most countable.

39. Let $\{\mathscr{A}_n\}$ be an increasing sequence of σ-algebras in Ω, and let $\mathscr{A} = \sigma\left(\bigcup_n \mathscr{A}_n\right)$. Assume that μ is a finite measure and ν is a probability measure on $(\Omega, \mathscr{A})$. Let μ_n, ν_n be the restrictions of μ, ν on $\mathscr{A}_n$, respectively. If $\mu_n \ll \nu_n$, $f = \overline{\lim_n} \frac{\mathrm{d}\mu_n}{\mathrm{d}\nu_n}$, prove

$$\mu(A) = \int_A f \,\mathrm{d}\nu + \mu(A \cap \{f = \infty\}), \quad A \in \mathscr{A}.$$

Chapter 4

Product Measure Space

Why should we study multi-dimensional spaces and even infinite-dimensional spaces? Let's consider a system of n many random particles in the 3-dimensional real world, where the location of each particle is a 3-dimensional random variable, and the joint distribution of these particles is a probability measure on the $3n$-dimensional Euclidean space. Another example is to consider a particle randomly moving on the real line, at each time its location is a 1-dimensional random variable. If we want to describe the movement of the particle, we have to clarify its path when time varies, which is an infinite-dimensional random variable (stochastic process), whose distribution is a probability measure on an infinite product space. Note that the finite product measure space has been introduced in Section 1.1.5 of Chapter 1 and Corollary 3.14 of Chapter 3. We will extend the notion to the infinite product case.

In this chapter, we first establish Fubini's theorem, which reduces the integral with respect to a product measure to the iterated integrals with respect to the marginal measures, as we have already studied in Lebesgue's measure theory. We then extend this theorem to nonproduct measures induced by a marginal measure and transition measures. In particular, the construction of probability measures on infinite product measurable spaces is fundamental in the study of stochastic processes.

4.1 Fubini's Theorem

Let us recall how to reduce a multiple integral on $\mathbb{R}^2$ to iterated integrals, as we learnt in calculus or Lebesgue's integral theory. Let A be a measurable subset of $\mathbb{R}^2$ and f be an integrable function with respect to the 2-dimensional Lebesgue measure. To compute $\int_A f(x_1, x_2)\,\mathrm{d}x_1\,\mathrm{d}x_2$, we first determine x_1 and fix the range of x_2, i.e. $A_{x_1} := \{x_2 \in \mathbb{R} : (x_1, x_2) \in A\}$. Then the multiple integral can be calculated as $\int_{\mathbb{R}} \mathrm{d}x_1 \int_{A_{x_1}} f(x_1, x_2)\,\mathrm{d}x_2$. The aim of this section is to realize this procedure for integrals on a general product measure space.

Let $(\Omega_1, \mathscr{A}_1, \mu_1)$ and $(\Omega_2, \mathscr{A}_2, \mu_2)$ be σ-finite measure spaces. By Corollary 3.14 of Chapter 3, the product measure space $(\Omega_1 \times \Omega_2, \mathscr{A}_1 \times \mathscr{A}_2, \mu_1 \times \mu_2)$ is σ-finite as well. Given $A \in \mathscr{A}_1 \times \mathscr{A}_2$ and a measurable function f which is integrable with respect to $\mu_1 \times \mu_2$, we will prove

$$\int_A f\,\mathrm{d}(\mu_1 \times \mu_2) = \int_{\Omega_1} \mu_1(\mathrm{d}\omega_1) \int_{A_{\omega_1}} f(\omega_1, \omega_2)\mu_2(\mathrm{d}\omega_2), \tag{4.1}$$

where A_{ω_1} is the section of A at ω_1. This formula is called Fubini's theorem. For this, we first introduce the section of a set.

Definition 4.1. Let $A \subset \Omega_1 \times \Omega_2$. $\forall \omega_1 \in \Omega_1$,

$$A_{\omega_1} := \{\omega_2 \in \Omega_2 : (\omega_1, \omega_2) \in A\} \subset \Omega_2$$

is called the section of set A at ω_1. Similarly, we can define $A_{\omega_2} \subset \Omega_1$, $\forall \omega_2 \in \Omega_2$.

Clearly, sections of sets have the following properties.

Property 4.2.

(1) $A \cap B = \varnothing \Rightarrow A_{\omega_i} \cap B_{\omega_i} = \varnothing$.
(2) $A \supset B \Rightarrow A_{\omega_i} \supset B_{\omega_i}$.
(3) $\left(\bigcup_n A^{(n)}\right)_{\omega_i} = \bigcup_n A^{(n)}_{\omega_i}$.
(4) $\left(\bigcap_n A^{(n)}\right)_{\omega_i} = \bigcap_n A^{(n)}_{\omega_i}$,
(5) $(A - B)_{\omega_i} = A_{\omega_i} - B_{\omega_i}$.

To prove (4.1), we need to clarify that the right-hand side of this formula makes sense by verifying that $\forall \omega_1 \in \Omega_1$, we have $A_{\omega_1} \in \mathscr{A}_2$,

$\int_{A_{\omega_1}} f(\omega_1, \omega_2)\mu_2(\mathrm{d}\omega_2)$ is $\mathscr{A}_1$-measurable in ω_1 and has integral with respect to μ_1.

Theorem 4.3. *Let $A \in \mathscr{A}_1 \times \mathscr{A}_2$. Then for any $\omega_i \in \Omega_i, i = 1, 2$, we have $A_{\omega_1} \in \mathscr{A}_2$ and $A_{\omega_2} \in \mathscr{A}_1$.*

Proof. Let

$$\mathscr{M} = \{A \in \mathscr{A}_1 \times \mathscr{A}_2 : \forall \omega_1 \in \Omega_1, A_{\omega_1} \in \mathscr{A}_2; \forall \omega_2 \in \Omega_2, A_{\omega_2} \in \mathscr{A}_1\}.$$

Clearly, $\mathscr{M}$ includes the semi-algebra $\{A_1 \times A_2 : A_1 \in \mathscr{A}_1, A_2 \in \mathscr{A}_2\}$. By Property 4.2, we see that $\mathscr{M}$ is a σ-algebra, so it includes $\mathscr{A}_1 \times \mathscr{A}_2$. □

Recall that for a function f and a σ-algebra $\mathscr{A}$, $f \in \mathscr{A}$ means that f is $\mathscr{A}$-measurable.

Theorem 4.4. *For any $\mathscr{A}_1 \times \mathscr{A}_2$-measurable function f and any $\omega_i \in \Omega_i, i = 1, 2$, we have $f_{\omega_1}(\cdot) := f(\omega_1, \cdot) \in \mathscr{A}_2$ and $f_{\omega_2}(\cdot) := f(\cdot, \omega_2) \in \mathscr{A}_1$.*

Proof. For any $B \in \mathscr{B}$, we have

$$\begin{aligned} f_{\omega_1}^{-1}(B) &= \{\omega_2 \in \Omega_2 : f_{\omega_1}(\omega_2) \in B\} \\ &= \left\{\omega_2 \in \Omega_2 : (\omega_1, \omega_2) \in f^{-1}(B)\right\} = \left[f^{-1}(B)\right]_{\omega_1}, \end{aligned}$$

which is in $\mathscr{A}_2$ by Theorem 4.3. So, f_{ω_1} is $\mathscr{A}_2$-measurable. Similarly, f_{ω_2} is $\mathscr{A}_1$-measurable. □

The functions f_{ω_1} and f_{ω_2} are called the section functions f at ω_1 and ω_2, respectively.

Theorem 4.5. *Let f be a nonnegative measurable function on $(\Omega_1 \times \Omega_2, \mathscr{A}_1 \times \mathscr{A}_2)$. Then*

$$\int_{\Omega_1} f(\omega_1, \omega_2)\mu_1(\mathrm{d}\omega_1) \in \mathscr{A}_2, \quad \int_{\Omega_2} f(\omega_1, \omega_2)\mu_2(\mathrm{d}\omega_2) \in \mathscr{A}_1.$$

Proof. By the construction of measurable functions and the properties of integrals, we only need to prove for $f = \mathbf{1}_A$ with $A \in \mathscr{A}_1 \times \mathscr{A}_2$, and by the monotone class theorem, it suffices to consider

$A = A_1 \times A_2, A_i \in \mathscr{A}_i, i = 1, 2$. In this case, we have

$$\int_{\Omega_1} f(\omega_1, \omega_2)\mu_1(\mathrm{d}\omega_1) = \mu_1(A_1)\mathbf{1}_{A_2} \in \mathscr{A}_2,$$

$$\int_{\Omega_2} f(\omega_1, \omega_2)\mu_2(\mathrm{d}\omega_2) = \mu_2(A_2)\mathbf{1}_{A_1} \in \mathscr{A}_1. \qquad \square$$

Theorem 4.6 (Fubini's theorem). *Let f be $\mathscr{A}_1 \times \mathscr{A}_2$-measurable function having integral with respect to $\mu_1 \times \mu_2$. Then*

$$\int_{\Omega_1 \times \Omega_2} f \,\mathrm{d}\mu_1 \times \mu_2 = \int_{\Omega_1} \left(\int_{\Omega_2} f(\omega_1, \omega_2)\mu_2(\mathrm{d}\omega_2) \right) \mu_1(\mathrm{d}\omega_1)$$
$$= \int_{\Omega_2} \left(\int_{\Omega_1} f(\omega_1, \omega_2)\mu_1(\mathrm{d}\omega_1) \right) \mu_2(\mathrm{d}\omega_2).$$

Proof. By symmetry, we only prove the first equation.

(1) The equation holds obviously for $f = \mathbf{1}_{A_1 \times A_2}$ $(A_i \in \mathscr{A}_i, i = 1, 2)$. From this, the monotone class theorem implies that the equation holds for $f = \mathbf{1}_A$ $(A \in \mathscr{A}_1 \times \mathscr{A}_2)$.
(2) By the linear property of integral and step (1) in the proof, we obtain the equation for a simple function f. Combining this with Theorem 2.12(4) of Chapter 2 and Theorem 3.5 of Chapter 3, we prove the equation for a nonnegative measurable function f.
(3) For a general measurable function f such that $(\mu_1 \times \mu_2)(f)$ exists, assume, for instance, $(\mu_1 \times \mu_2)(f^-) < \infty$. By step (2), we have

$$\infty > \int_{\Omega_1 \times \Omega_2} f^- \,\mathrm{d}(\mu_1 \times \mu_2) = \int_{\Omega_1} \mu_1(\mathrm{d}\omega_1) \int_{\Omega_2} f^-(\omega_1, \cdot) \,\mathrm{d}\mu_2.$$

Thus, μ_1-a.e. ω_1, $\int_{\Omega_2} f^-(\omega_1, \cdot) \,\mathrm{d}\mu_2 < \infty$, so that

$$\int_{\Omega_2} f(\omega_1, \cdot) \,\mathrm{d}\mu_2 = \int_{\Omega_2} f^+(\omega_1, \cdot) \,\mathrm{d}\mu_2 - \int_{\Omega_2} f^-(\omega_1, \cdot) \,\mathrm{d}\mu_2.$$

Combining this with the linear property of integral and step (2), we finish the proof. $\square$

By applying Theorem 4.6 to $f\mathbf{1}_A$ in place of f, we derive (4.1). Moreover, by induction, Fubini's theorem can be extended to multi-product measure spaces.

Let $(\Omega_i, \mathscr{A}_i, \mu_i), 1 \leqslant i \leqslant n$ be σ-finite measure spaces and f be a measurable function having integral on the product measure space $(\Omega, \mathscr{A}, \mu) := (\Omega_1 \times \cdots \times \Omega_n, \mathscr{A}_1 \times \cdots \times \mathscr{A}_n, \mu_1 \times \cdots \times \mu_n)$. Then

$$\int_\Omega f\, \mathrm{d}\mu = \int_{\Omega_{i_1}} \mathrm{d}\mu_{i_1} \int_{\Omega_{i_2}} \mathrm{d}\mu_{i_2} \cdots \int_{\Omega_{i_n}} f\, \mathrm{d}\mu_{i_n},$$

where $(i_1, \ldots, i_n)$ is any permutation of $(1, \ldots, n)$. This means that all integrals in the right-hand side exist, and the iterated integral equals to the multiple integral in the left-hand side.

4.2 Infinite Product Probability Space

Let $\{(\Omega_t, \mathscr{A}_t, \mathbb{P}_t)\}_{t\in T}$ be a family of probability spaces, where T is an infinite index set. Let

$$\Omega_T = \prod_{t\in T} \Omega_t = \{\omega : \omega = (\omega_t)_{t\in T}, \omega_t \in \Omega_t, t \in T\}.$$

We intend to define the produce σ-algebra $\mathscr{A}_T = \prod \pi_{t\in T}\mathscr{A}_t$ and the product probability measure $\mathbb{P}_T = \prod_{t\in T} \mathbb{P}_t$.

Following the line in the finite product setting, one may define $\mathscr{A}_T$ as the σ-algebra generated by the class of rectangles $\{\prod_{t\in T} A_t : A_t \in \mathscr{A}_t, t \in T\}$. However, this class is not a semi-algebra as required by the measure extension theorem, and for each rectangle $\prod_{t\in T} A_t$, its probability $\prod_{t\in T} \mathbb{P}_t(A_t)$ is usually ill-defined. For this reason, we only allow the set $\{\mathbb{P}_t(A_t) : t \in T\}$ to be finite, so a natural way is to restrict ourselves to the following class of measurable cylindrical sets. This also explains why we only study infinite product probability measures rather than infinite product measures.

Definition 4.7. A set like

$$B^{T_N} \times \prod_{t\notin T_N} \Omega_t$$

is called a measurable cylindrical set, where $T_N \Subset T$ (i.e. finite subset), and $B^{T_N} \in \mathscr{A}_{T_N} := \prod_{t\in T_N} \mathscr{A}_t$. In this case, B^{T_N} is called the base of the cylindrical set (it is not unique!).

Let $\mathscr{A}^T$ be the total of measurable cylindrical sets, which is obviously an algebra. Define infinite product σ-algebra by

$$\mathscr{A}_T = \prod_{t\in T} \mathscr{A}_t := \sigma(\mathscr{A}^T).$$

Theorem 4.8. *There exists a unique probability $\mathbb{P}$ on $(\Omega_T, \mathscr{A}_T)$ such that*

$$\mathbb{P}\left(A^{T_N} \times \Omega_{T_N^c}\right) = \left(\prod_{t\in T_N} \mathbb{P}_t\right)(A^{T_N}), \tag{4.2}$$

where

$$T_N \Subset T, \quad \Omega_{T_N^c} = \prod_{t\notin T_N} \Omega_t, \quad A^{T_N} \in \mathscr{A}_{T_N}.$$

Proof. (1) Formula (4.2) defines a function on $\mathscr{A}^T$, so we need to prove this function of sets is independent of the choice of repressions of the cylindrical set.

Let $A^{T_N} \times \Omega_{T_N^c} = A^{T_N'} \times \Omega_{(T_N')^c}$ with $T_N, T_N' \Subset T$, $A^{T_N} \in \mathscr{A}_{T_N}$, $A^{T_N'} \in \mathscr{A}_{T_N'}$. Let $T_N'' = T_N \cap T_N'$. Then there exists $A^{T_N''} \in \prod_{t\in T_N''} \mathscr{A}_t$ such that

$$A^{T_N} = A^{T_N''} \times \prod_{t\in T_N - T_N''} \Omega_t, A^{T_N'} = A^{T_N''} \times \prod_{t\in T_N' - T_N''} \Omega_t.$$

Hence, $\left(\prod_{t\in T_N} \mathbb{P}_t\right)(A^{T_N}) = \left(\prod_{t\in T_N''} \mathbb{P}_t\right)(A^{T_N''}) = \left(\prod_{t\in T_N'} \mathbb{P}_t\right)(A^{T_N'})$.

(2) $\mathbb{P}$ is finitely additive.

Assume $\{A_k\}_{k=1}^n \subset \mathscr{A}^T$ are mutually disjoint. For $1 \leqslant k \leqslant n$, let T_k be a finite subset of T and $A^{T_k} \in \prod_{t\in T_k} \mathscr{A}_t$ such that $A_k = A^{T_k} \times \Omega_{T_k^c}$. Then $\sum_{k=1}^n A_k =: A_0 \in \mathscr{A}^T$. Let $T_0 \subset T$ be a finite set and $A^{T_0} \in \prod_{t\in T_0} \mathscr{A}_t$ such that $A_0 = A^{T_0} \times \Omega_{T_0^c}$.

Set $T_N = \cup_{k=0}^n T_k, A_k^{T_N} = A^{T_k} \times \prod_{t\in T_N - T_k} \Omega_t$. Then $A_k = A_k^{T_N} \times \Omega_{(T_N)^c}, n \geqslant k \geqslant 0$. Clearly, $\left\{A_k^{T_N}\right\}_{k=1}^n \subset \prod_{t\in T_N} \mathscr{A}_t$ are mutually disjoint, and $\sum_{k=1}^n A_k^{T_N} = A_0$. For any finite $T' \subset T$, let

$\mathbb{P}_{T'} = \prod_{t\in T'} \mathbb{P}_t$ be a product measure on $(\Omega_{T'}, \mathscr{A}_{T'})$, but $\mathbb{P}^{T'}$ is a finitely additive function on $\mathscr{A}^{T\setminus T'}$, defined by (4.2) with $T \setminus T'$ as its total index set. Since $\mathbb{P}_{T_N} := \prod_{t\in T_N} \mathbb{P}_t$ is a probability measure, it follows from the definition of $\mathbb{P}$ and the finite additivity of measure that $\sum_{k=1}^{n} \mathbb{P}(A_k) = \mathbb{P}(A_0)$.

(3) Since $\mathscr{A}^T$ is a set algebra and $\mathbb{P}$ is finitely additive, to get the σ-additivity of $\mathbb{P}$, we only need to prove that it is continuous at $\varnothing$. We use the method of proof by contradiction.

Let $\{A_n\}_{n\geqslant 1} \subset \mathscr{A}^T$ be decreasing and $\exists \varepsilon > 0$ such that $\mathbb{P}(A_n) \geqslant \varepsilon$ for every $n \geqslant 1$. Now, we prove $\cap_{n=1}^{\infty} A_n \neq \varnothing$. Note that for any $n \geqslant 1$, there exist a finite set $T_n \subset T$ and $A_n^{T_n} \in \prod_{t\in T_n} \mathscr{A}_t$ such that $A_n = A_n^{T_n} \times \prod_{t\notin T_n} \Omega_t$.

Let $T_\infty = \cup_{n=1}^{\infty} T_n$. Then T_∞ is countable, denoted by $T_\infty = \{t_1, t_2, \ldots\}$. To prove $\cap_{n=1}^{\infty} A_n \neq \varnothing$, we only need to prove $\exists(\bar{\omega}_{t_1}, \ldots, \bar{\omega}_{t_n}, \ldots) \in \prod_{t\in T_\infty} \Omega_t$, such that $\bigcap_{j=1}^{\infty} A_j(\bar{\omega}_{t_1}, \ldots, \bar{\omega}_{t_n}, \ldots) \neq \varnothing$, where $A_j(\bar{\omega}_{t_1}, \ldots, \bar{\omega}_{t_n}, \ldots)$ is section of A_j at $(\bar{\omega}_{t_1}, \ldots, \bar{\omega}_{t_n}, \ldots)$.

First, we set $B_1^{(j)} = \left\{\omega_{t_1} \in \Omega_{t_1} : \mathbb{P}^{\{t_1\}}(A_j(\omega_{t_1})) \geqslant \frac{\varepsilon}{2}\right\}$. Since

$$\mathbb{P}^{\{t_1\}}(A_j(\omega_{t_1})) = \mathbb{P}_{t\in T_j\setminus\{t_1\}} \left(A_j^{T_j}(\omega_{t_1})\right)$$

is $\mathscr{A}_{t_1}$-measurable, $B_1^{(j)} \in \mathscr{A}_{t_1}$. Fubini's theorem gives

$$\varepsilon \leqslant \mathbb{P}(A_j) = \int_{\Omega_{t_1}} \mathbb{P}^{\{t_1\}}\left(A_j(\omega_{t_1})\right) \mathrm{d}\mathbb{P}_{t_1} \leqslant \mathbb{P}_{t_1}\left(B_1^{(j)}\right) + \frac{\varepsilon}{2},$$

which implies that $\mathbb{P}_{t_1}\left(B_1^{(j)}\right) \geqslant \frac{\varepsilon}{2}$. Since $\left\{B_1^{(j)}\right\}_{j=1}^{\infty}$ is decreasing, we have $\mathbb{P}_{t_1}\left(\bigcap_{j=1}^{\infty} B_1^{(j)}\right) \geqslant \frac{\varepsilon}{2}$, so $\exists \bar{\omega}_{t_1} \in \bigcap_{j=1}^{\infty} B_1^{(j)}$, that is, $\mathbb{P}^{\{t_i\}}(A_j(\tilde{\omega}_{t_j})) \geqslant \frac{\varepsilon}{2}$ for every $j \geqslant 1$.

In general, assume for some $k \geqslant 1$ we have $(\bar{\omega}_{t_1}, \ldots, \bar{\omega}_{t_k}) \in \Omega_{t_1} \times \cdots \Omega_{t_k}$ such that

$$\mathbb{P}^{\{t_1,\ldots,t_k\}}(A_j(\bar{\omega}_{t_1}, \ldots, \bar{\omega}_{t_k})) \geqslant \frac{\varepsilon}{2^k}, \quad \forall j \geqslant 1.$$

Let

$$B_{k+1}^{(j)} = \Big\{\omega_{t_{k+1}} \in \Omega_{t_{k+1}} : \mathbb{P}^{\{t_1,\ldots,t_k,t_{k+1}\}}(A_j(\bar{\omega}_{t_1}, \ldots, \bar{\omega}_{t_k}, \omega_{t_{k+1}})) \geqslant \frac{\varepsilon}{2^{k+1}}\Big\}.$$

Then it follows from Fubini's theorem and induction that

$$\begin{aligned}\frac{\varepsilon}{2^k} &\leqslant \mathbb{P}^{\{t_1,\ldots,t_k\}}(A_j(\bar{\omega}_{t_1},\ldots,\bar{\omega}_{t_k}))\\ &= \int_{\Omega_{t_{k+1}}} \mathbb{P}^{\{t_1,\ldots,t_k,t_{k+1}\}}(A_j(\bar{\omega}_{t_1},\ldots,\bar{\omega}_{t_k},\omega_{t_{k+1}}))\,\mathrm{d}\mathbb{P}_{t_{k+1}}(\omega_{t_{k+1}})\\ &\leqslant \mathbb{P}\left(B_{k+1}^{(j)}\right)+\frac{\varepsilon}{2^{k+1}}.\end{aligned}$$

Thus, $\mathbb{P}\left(B_{k+1}^{(j)}\right) \geqslant \frac{\varepsilon}{2^{k+1}}$ for every $j \geqslant 1$. Hence, $\exists \bar{\omega}_{t_{k+1}} \in \bigcap_{j=1}^{\infty} B_{k+1}^{(j)}$, i.e.

$$\mathbb{P}^{\{t_1,\ldots,t_k,t_{k+1}\}}(A_j(\bar{\omega}_{t_1},\ldots,\bar{\omega}_{t_k},\bar{\omega}_{t_{k+1}})) \geqslant \frac{\varepsilon}{2^{k+1}},\ \forall j \geqslant 1.$$

By induction, it follows that $\exists \{\bar{\omega}_{t_i} \in \Omega_{t_i}\}_{i\geqslant 1}$ such that $\bigcap_{j=1}^{\infty} A_j(\bar{\omega}_{t_1},\ldots,\bar{\omega}_{t_n}) \neq \varnothing$ for every $n \geqslant 1$.

Take and fix $\tilde{\omega} \in \prod_{t\notin T_\infty} \Omega_t$ and let $\omega \in \prod_{t\in T} \Omega_t$ such that

$$\omega_t = \begin{cases} \bar{\omega}_t, & \text{for } t \in T_\infty,\\ \tilde{\omega}_t, & \text{for } t \notin T_\infty. \end{cases}$$

Then for any $j \geqslant 1$, there exists N_j such that $T_j \subset \{t_1,\ldots,t_{N_j}\}$, but $A_j(\bar{\omega}_{t_1},\ldots,\bar{\omega}_{t_{N_j}}) \neq \varnothing$, so $\omega \in A_j$ for every $j \geqslant 1$. Hence, $\omega \in \bigcap_j A_j$. □

Theorem 4.9. *Let $\{F_t\}_{t\in T}$ be a family of probability distributions. Then there exists a family of independent random variables $\{\xi_t\}_{t\in T}$ such that ξ_t has distribution function F_t for each $t \in T$.*

Proof. Assume $\mathbb{P}_t$ is probability measure induced by F_t on $(\mathbb{R},\mathscr{B})$. Let

$$(\Omega_t,\mathscr{A}_t,\mathbb{P}_t) = (\mathbb{R},\mathscr{B},\mathbb{P}_t),\ \Omega = \mathbb{R}^T,\ \mathscr{A} = \mathscr{B}^T,\ \mathbb{P} = \prod_{t\in T}\mathbb{P}_t.$$

Then $(\xi_t(\omega) := \omega_t)_{t\in T}$ are random variables on $(\Omega,\mathscr{A},\mathbb{P})$, and

$$\mathbb{P}(\xi_t < x_t) = \mathbb{P}_t((-\infty,x_t)) = F_t(x_t).$$

Obviously, they are independent. □

4.3 Transition Measure and Transition Probability

As shown in Theorem 4.9, the joint distribution of independent random variables is a product measure. In this section, we aim to construct non-product measures on a product measurable space. To this end, we introduce the notion of transition measure, in particular transition probability, which describes the conditional distribution of a random variable given the value of another random variable. See Chapter 5 for details.

Definition 4.10. Let $(\Omega_i, \mathscr{A}_i)(i = 1, 2)$ be two measurable spaces. A map $\lambda : \Omega_1 \times \mathscr{A}_2 \to [0, \infty]$ is called a transition measure from $(\Omega_1, \mathscr{A}_1)$ to $(\Omega_2, \mathscr{A}_2)$ or simply a transition measure on $\Omega_1 \times \mathscr{A}_2$ if it has the following two properties:

(1) $\lambda(\omega_1, \cdot)$ is a measure on $(\Omega_2, \mathscr{A}_2)$ for any $\omega_1 \in \Omega_1$,
(2) $\lambda(\cdot, A)$ is a measurable function of $\mathscr{A}_1$ for any $A \in \mathscr{A}_2$.

If there exists a partition $\{B_n\}_{n\in\mathbb{N}} \subset \mathscr{A}_2$ of Ω_2 such that $\lambda(\omega_1, B_n) < \infty\,(n \geqslant 1,\ \omega_1 \in \Omega_1)$, then λ is called σ-finite. Furthermore, if $\sup_{\omega_1\in\Omega_1} \lambda(\omega_1, B_n) < \infty\,(\forall n \geqslant 1)$, λ is called uniformly σ-finite. If $\lambda(\omega_1, \cdot)$ is a probability for any $\omega_1 \in \Omega_1$, then λ is called a transition probability.

To construct non-product measures on a product space by using transition measures, and to extend Fubini's theorem for the integral with respect to such a measure, we need the following theorem.

Theorem 4.11. *Let λ be a σ-finite transition measure on $\Omega_1 \times \mathscr{A}_2$, and let f be a nonnegative measurable function on $\mathscr{A}_1 \times \mathscr{A}_2$. Then $\int_{\Omega_2} f(\cdot, \omega_2)\lambda(\cdot,\, d\omega_2)$ is $\mathscr{A}_1$-measurable.*

Proof. By Theorem 2.12 of Chapter 2 and Theorem 3.8 of Chapter 3, we only need to prove for f being an indicator function. Moreover, by the monotone class theorem, it suffices to consider $f = \mathbf{1}_{A\times B}$ for some $A \in \mathscr{A}_1, B \in \mathscr{A}_2$. In this case, $\int_{\Omega_2} f(\cdot, \omega_2)\lambda(\cdot,\, d\omega_2) = \lambda(\cdot, B)\mathbf{1}_A$, which is obviously $\mathscr{A}_1$-measurable. □

Theorem 4.12. *Let* $(\Omega_i, \mathscr{A}_i)(i = 1, 2, \ldots, n)$ *be finite many measurable spaces, and let*

$$\Omega^{(k)} = \prod_{i=1}^{k} \Omega_i, \quad \mathscr{A}^{(k)} = \prod_{i=1}^{k} \mathscr{A}_i, \quad k = 1, \ldots, n.$$

If λ_1 *is a* σ*-finite measure on* $\mathscr{A}_1$ *and* λ_k *is a* σ*-finite transition measure on* $\Omega^{(k-1)} \times \mathscr{A}_k$ *for each* $k = 2, \ldots, n$*, then*

$$\lambda^{(n)}(B) := \int_{\Omega_1} \cdots \int_{\Omega_n} 1_B(\omega_1, \ldots, \omega_n)\lambda_n (\omega_1, \ldots, \omega_{n-1}, \mathrm{d}\omega_n) \ldots \lambda_1(\mathrm{d}\omega_1)$$

for any $B \in \mathscr{A}^{(n)}$ *defines a measure on* $\mathscr{A}^{(n)}$*. If* $\{\lambda_i\}_{i=2,\cdots n}$ *are uniformly* σ*-finite, then* $\lambda^{(n)}$ *is* σ*-finite.*

Proof. By Theorem 4.11, $\lambda^{(n)}$ is a well-defined nonnegative function on $\mathscr{A}^{(n)}$. By applying Corollary 3.12 for n many times, we see that $\lambda^{(n)}$ is σ-additive. Hence, it is a measure on $\mathscr{A}^{(n)}$.

Now, let $\{\lambda_i\}_{i=2,\ldots,n}$ be uniformly σ-finite, we intend to prove that $\lambda^{(n)}$ is σ-finite. By induction, we only prove for $n = 2$.

Since λ_1 is σ-finite and λ_2 is uniformly σ-finite, we find measurable partitions $\{A_n\}_{n\in\mathbb{N}}$ for Ω_1 and $\{B_n\}_{n\in\mathbb{N}}$ for Ω_2 such that $\lambda_1(A_n) < \infty$ and $\sup_{\omega_1 \in A_m} \lambda_2(\omega_1, B_n) < \infty$ for all $m, n \geqslant 1$. Then $\{A_i \times B_j\}_{i,j\geqslant 1}$ is a measurable partition of $\Omega_1 \times \Omega_2$ satisfying

$$\begin{aligned}\lambda^{(2)}(A_i \times B_j) &= \int_{A_i} \lambda_1(\mathrm{d}\omega_1) \int_{B_j} \lambda_2(\omega_1, \mathrm{d}\omega_2) \\ &= \int_{A_i} \lambda_2(\omega_1, B_j)\lambda_1(\mathrm{d}\omega_1) \\ &\leqslant \sup_{\omega_1 \in A_i} \lambda_2(\omega_1, B_j)\lambda_1(A_i) < \infty.\end{aligned}$$

□

Theorem 4.13 (Generalized Fubini's theorem). *Let* $\Omega^{(n)}$, $\mathscr{A}^{(n)}$ *and* $\lambda^{(n)}$ *be in Theorem 4.12, and let* f *be a measurable function on* $(\Omega^{(n)}, \mathscr{A}^{(n)})$*. Then the integral* $\lambda^{(n)}(f)$ *exists if and only if at least* *one of* $I(f^+)$ *and* $I(f^-)$ *is finite, where*

$$I(f^{\pm}) := \int_{\Omega_1} \cdots \int_{\Omega_n} f^{\pm}(\omega_1, \ldots, \omega_n)$$
$$\lambda_n(\omega_1, \ldots, \omega_{n-1}, \mathrm{d}\omega_n) \cdots \lambda_1(\mathrm{d}\omega_1),$$

and in this case we have

$$\lambda^{(n)}(f) = \int_{\Omega_1} \cdots \int_{\Omega_n} f(\omega_1, ..., \omega_n)\lambda_n(\omega_1, \ldots, \omega_{n-1}, \mathrm{d}\omega_n) \cdots \lambda_1(\mathrm{d}\omega_1).$$

Proof. When $f = \mathbf{1}_B, B \in \mathscr{A}^{(n)}$, the formula follows from Theorem 4.12. Combining this with Theorem 2.12 of Chapter 2 and Theorem 3.8 of Chapter 3, we prove the formula for all any nonnegative measurable function f. In general, by the definition of integral, by the formula for nonnegative functions, we have $\lambda^{(n)}(f^{\pm}) = I(f^{\pm})$, so that $\lambda^{(n)}(f)$ exists if and only if at least one of $I(f^+)$ and $I(f^-)$ is finite, and in this case,

$$\begin{aligned}
&\int f \, \mathrm{d}\lambda^{(n)} = \int f^+ \, \mathrm{d}\lambda^{(n)} - \int f^- \, \mathrm{d}\lambda^{(n)} = I(f^+) - I(f^-) \\
&= \int_{\Omega_1} \mathrm{d}\lambda_1 \left\{ \int_{\Omega_2} \cdots \int_{\Omega_n} f^+ \, \mathrm{d}\lambda_n \cdots \mathrm{d}\lambda_2 - \int_{\Omega_2} \cdots \int_{\Omega_n} f^- \, \mathrm{d}\lambda_n ... \mathrm{d}\lambda_2 \right\} \\
&= \cdots = \int_{\Omega_1} \cdots \int_{\Omega_n} (f^+ - f^-) \, \mathrm{d}\lambda_n \cdots \mathrm{d}\lambda_1 \\
&= \int_{\Omega_1} \cdots \int_{\Omega_n} f \, \mathrm{d}\lambda_n \cdots \mathrm{d}\lambda_1.
\end{aligned}$$

□

Finally, we construct probability measures on an infinite product measurable space by using a marginal distribution $\mathbb{P}_1$ and a sequence of transition probability measures $\{\mathbb{P}_n\}_{n \geqslant 1}$.

Theorem 4.14 (Tulcea's theorem). *Let* $(\Omega_n, \mathscr{A}_n)_{n \in \mathbb{N}}$ *be a sequence of measurable spaces, and let* $\left(\Omega^{(n)}, \mathscr{A}^{(n)}\right)$ *be defined in Theorem* 4.12. *Set*

$$\Omega = \prod_{i=1}^{\infty} \Omega_i, \quad \mathscr{A} = \prod_{i=1}^{\infty} \mathscr{A}_i.$$

Let $\mathbb{P}_1$ *be a probability measure on* $(\Omega_1, \mathscr{A}_1)$, *and for each* $n \geqslant 2$, *let* $\mathbb{P}_n$ *be a transition probability on* $\Omega^{(n-1)} \times \mathscr{A}_n$. *Then there exists a unique probability measure* $\mathbb{P}$ *on* $(\Omega, \mathscr{A})$ *such that*

$$\mathbb{P}\left(B^{(n)} \times \prod_{k>n} \Omega_k\right) = \mathbb{P}^{(n)}, \quad n \in \mathbb{N}, \quad B^{(n)} \in \mathscr{A}^{(n)},$$

where $\mathbb{P}^{(n)}\left(B^{(n)}\right) = \int_{\Omega_1} \cdots \int_{\Omega_n} \mathbf{1}_{B^{(n)}} \, \mathrm{d}\mathbb{P}_n \cdots \mathrm{d}\mathbb{P}_1$.

Proof. Let $\mathscr{C}^{(\infty)}$ be class of all measurable cylindrical sets, which is an algebra in Ω. As explained in the proof of Theorem 4.8, $\mathbb{P}$ is a well-defined finitely additive nonnegative function on $\mathscr{C}^{(\infty)}$ with $\mathbb{P}(\Omega) = 1$. Moreover, by using Theorem 4.13 in place of Fubini's theorem, the same argument in the proof of Theorem 4.8 implies the continuity of $\mathbb{P}$ at $\varnothing$. Hence, $\mathbb{P}^{(\infty)}$ is a probability on $\mathscr{C}^{(\infty)}$, which is uniquely extended to a probability on $(\Omega, \mathscr{A})$. □

4.4 Exercises

1. Prove Property 4.2.
2. Let $(\Omega, \mathscr{A}, \mu)$ be a measure space and let f be a nonnegative measurable function. Prove

$$\mu(f) = \int_0^\infty \mu(f > r) \, \mathrm{d}r = \int_0^\infty \mu(f \geqslant r) \, \mathrm{d}r.$$

3. Let $(\Omega_i, \mathscr{A}_i, \mu_i)(i = 1, 2)$ be measure spaces, and let $A, B \in \mathscr{A}_1 \times \mathscr{A}_2$. If $\mu_2 \{\omega_2 : (\omega_1, \omega_2) \in A\} = \mu_2 \{\omega_2 : (\omega_1, \omega_2) \in B\}$ holds for μ_1-a.e. ω_1, prove $(\mu_1 \times \mu_2)(A) = (\mu_1 \times \mu_2)(B)$.
4. Let $(\Omega_i, \mathscr{A}_i)(i=1,2,3)$ be measurable spaces, λ be a σ-finite transition measure on $\Omega_2 \times \mathscr{A}_3$, and f be a measurable function on $(\Omega_1 \times \Omega_3, \mathscr{A}_1 \times \mathscr{A}_3$. If the integral $g(\omega_1, \omega_2) := \int_{\Omega_3} f(\omega_1, \omega_3) \lambda(\omega_2, \mathrm{d}\omega_3)$ exists for all $(\omega_1, \omega_2) \in \Omega_1 \times \Omega_2$, prove g is $\mathscr{A}_1 \times \mathscr{A}_2$-measurable.
5. Let $(\Omega_i, \mathscr{A}_i, \mu_i)(i = 1, 2)$ be σ-finite measure spaces. Prove that for any $A \in \mathscr{A}_1 \times \mathscr{A}_2$, the following statements are equivalent:

(a) $\mu_1 \times \mu_2(A) = 0$.
(b) $\mu_1(A_{\omega_2}) = 0$, μ_2-a.e.
(c) $\mu_2(A_{\omega_1}) = 0$, μ_1-a.e.

6. If an infinite matrix $P = (p_{ij})_{i,j\in\mathbb{N}}$ satisfies $p_{ij} \geqslant 0, \sum_{j\in\mathbb{N}} p_{ij} = 1, \forall i \in \mathbb{N}$, then P is called a transition probability matrix. Let $\lambda(i, A) = \sum_{j\in A} p_{ij}$, $i \in \mathbb{N}, A \subset \mathbb{N}$. Prove that λ is a transition probability on $\mathbb{N} \times 2^{\mathbb{N}}$.

7. Let μ be the counting measure on $\mathbb{N}$, i.e. $\mu(\{i\}) = 1$ for any $i \in \mathbb{N}$. Let

$$f(i,j) = \begin{cases} i, & i = j, \\ -i, & j = i+1, \\ 0, & \text{other } i, j. \end{cases}$$

Prove

$$\int_{\mathbb{N}} \left(\int_{\mathbb{N}} f(\omega_1, \omega_2) \mu(\mathrm{d}\omega_2) \right) \mu(\mathrm{d}\omega_1) = 0, \text{ but}$$

$$\int_{\mathbb{N}} \left(\int_{\mathbb{N}} f(\omega_1, \omega_2) \mu(\mathrm{d}\omega_1) \right) \mu(\mathrm{d}\omega_2) = \infty.$$

Does this contradict Fubini's theorem?

8. Construct a function $f : [0,1] \times [0,1] \to [0,1]$ fulfilling the following conditions:
   (a) $\forall z \in [0,1]$, the functions $f(z,\cdot)$ and $f(\cdot,z)$ are Borel measurable on $[0,1]$,
   (b) f is not Borel measurable on $[0,1] \times [0,1]$,
   (c) Both $\int_0^1 \left(\int_0^1 f(x,y)\,\mathrm{d}y\right) \mathrm{d}x$ and $\int_0^1 \left(\int_0^1 f(x,y)\,\mathrm{d}x\right) \mathrm{d}y$ exist, but do not equal.

9. Let μ_k, ν_k be σ-finite measures on $(\Omega_k, \mathscr{A}_k)$, respectively, and $\nu_k \ll \mu_k$ $(k = 1, 2)$. Prove that $\nu_1 \times \nu_2 \ll \mu_1 \times \mu_2$ and

$$\frac{\mathrm{d}(\nu_1 \times \nu_2)}{\mathrm{d}(\mu_1 \times \mu_2)}(\omega_1, \omega_2) = \frac{\mathrm{d}\nu_1}{\mathrm{d}\mu_1}(\omega_1) \frac{\mathrm{d}\nu_2}{\mathrm{d}\mu_2}(\omega_2), \ \mu_1 \times \mu_2\text{-a.e.}$$

10. Let $(\Omega_t, \mathscr{A}_t)_{t\in T}$ be a family of measurable spaces, where $\mathscr{A}_t = \sigma(\mathscr{C}_t)$ for $t \in T$. For each $t \in T$, let $\pi_t : \prod_{t\in T}\Omega_t \ni \omega \mapsto \omega_t \in \Omega_t$ be the projection onto the tth space. Prove $\prod_{t\in T}\mathscr{A}_t = \sigma\left(\bigcup_{t\in T}\pi_t^{-1}(\mathscr{C}_t)\right)$.

11. Let $\mathscr{B}^\infty$ be the product Borel σ-algebra on $\mathbb{R}^\infty := \prod_{i\in\mathbb{N}}\mathbb{R}$. Prove that the following sets are $\mathscr{B}^\infty$-measurable:

    (a) $\left\{x \in \mathbb{R}^\infty : \sup\limits_n x_n < a\right\}$,

    (b) $\left\{x \in \mathbb{R}^\infty : \sum\limits_{n=1}^{\infty} |x_n| < \infty\right\}$,

    (c) $\left\{x \in \mathbb{R}^\infty : \lim\limits_{n\to\infty} x_n \text{ exists and is finite}\right\}$,

    (d) $\left\{x \in \mathbb{R}^\infty : \overline{\lim\limits_n}\, x_n \leqslant a\right\}$.

12. Let F be a probability distribution function on $\mathbb{R}$. Prove $\int_{\mathbb{R}}(F(x+c) - F(x))\,\mathrm{d}x = c$ for any constant $c \in \mathbb{R}$, and if F is continuous, then $\int_{\mathbb{R}} F(x)\,\mathrm{d}F(x) = \frac{1}{2}$.

Chapter 5

Conditional Expectation and Conditional Probability

To describe the influence of a class of events (sub σ-algebra) $\mathscr{C}$ to a random variable ξ, we introduce the conditional expectation (or more generally, conditional distribution) of ξ given $\mathscr{C}$. When the sub σ-algebra $\mathscr{C}$ is induced by a family of random variables, the conditional expectation refers to the influence of these random variables to ξ. When ξ runs over the indicator functions for all measurable sets, the conditional expectation reduces to the conditional probability.

To define the conditional expectation, we recall the simple case where the condition is given by an event B. Throughout this chapter, $(\Omega, \mathscr{A}, \mathbb{P})$ is a complete probability space. For any $B \in \mathscr{A}$ with $\mathbb{P}(B) > 0$, the conditional probability given B is defined as $\mathbb{P}(\cdot|B) := \frac{\mathbb{P}(\cdot \cap B)}{\mathbb{P}(B)}$. Moreover, for any random variable ξ having expectation, the conditional expectation $\mathbb{E}(\xi|B)$ of a random variable ξ given B is defined as the integral of ξ with respect to conditional probability. Similarly, if $\mathbb{P}(B^c) > 0$, we define in the same way the conditional expectation $\mathbb{E}(\cdot|B^c)$ under event B^c. Thus, the conditional expectation of ξ given $\mathscr{C} = \{B, B^c, \varnothing, \Omega\}$ is naturally defined as

$$\mathbb{E}(\xi|\mathscr{C}) = \mathbf{1}_B \mathbb{E}(\xi|B) + \mathbf{1}_{B^c} \mathbb{E}(\xi|B^c), \tag{5.1}$$

which is a $\mathscr{C}$-measurable random variable. This is $\mathbb{P}|_{\mathscr{C}}$-a.s. well defined even if B or B^c is a $\mathbb{P}$-null set.

The aim of this chapter is to define the conditional probability and conditional expectation under an arbitrarily given sub-σ-algebra

$\mathscr{C}$ of $\mathscr{A}$ and make applications to the study of transition probabilities and probability measures on product spaces.

5.1 Conditional Expectation Given σ-Algebra

We first extend the definition in (5.1) to a σ-algebra $\mathscr{C}$ generated by countable many atoms. A set $B \in \mathscr{C}$ is called an atom of $\mathscr{C}$ if $\forall B' \in \mathscr{C}, B' \subset B$, we have $B' = B$ or $B' = \varnothing$.

Definition 5.1. Let $\mathscr{C} = \sigma(\{B_n : n \geqslant 1\})$ for $\{B_n\}_{n\geqslant 1} \subset \mathscr{A}$ being a partition of Ω, and let ξ be a random variable having expectation. Then

$$\mathbb{E}(\xi|\mathscr{C}) := \sum_{n=1}^{\infty} \mathbb{E}(\xi|B_n)\mathbf{1}_{B_n}$$

is called the conditional expectation of ξ with respect to $\mathbb{P}$ given σ-algebra $\mathscr{C}$, where $\mathbb{E}(\xi|B_n)\mathbf{1}_{B_n} := 0$ if $\mathbb{P}(B_n) = 0$.

To further extend the definition to general sub-σ-algebra $\mathscr{C}$, we present the following result which characterizes the conditional expectation without using the expression of $\mathscr{C}$.

Proposition 5.2. *Let $\mathscr{C} = \sigma(\{B_n : n \geqslant 1\})$ for a partition $\{B_n\}_{n\geqslant 1} \subset \mathscr{A}$ of Ω. Then for any random variable ξ having expectation, $\mathbb{E}(\xi|\mathscr{C})$ is $\mathscr{C}$-measurable and satisfies*

$$\mathbb{E}(1_B\xi) = \mathbb{E}\big[1_B\mathbb{E}(\xi|\mathscr{C})\big], \quad \forall B \in \mathscr{C}.$$

On the other hand, if η is a $\mathscr{C}$-measurable function such that $\mathbb{E}(\xi\mathbf{1}_B) = \mathbb{E}(\eta\mathbf{1}_B), \forall B \in \mathscr{C}$, then $\eta = \mathbb{E}(\xi|\mathscr{C})$, $\mathbb{P}|_{\mathscr{C}}$-a.s.

According to Proposition 5.2, we define the conditional expectation given a general σ-algebra $\mathscr{C}$ as follows.

Definition 5.3. Let $\mathscr{C} \subset \mathscr{A}$ be a sub-σ-algebra of $\mathscr{A}$, and let ξ be a random variable having expectation. The conditional expectation $\mathbb{E}(\xi|\mathscr{C})$ of ξ given $\mathscr{C}$ (with respect to $\mathbb{P}$) is defined as the $\mathscr{C}$-measurable function, satisfying

$$\int_B \mathbb{E}(\xi|\mathscr{C})\,\mathrm{d}\mathbb{P} = \int_B \xi\,\mathrm{d}\mathbb{P}, \quad \forall B \in \mathscr{C}.$$

To see that Definition 5.3 makes sense, we need to show the existence and uniqueness of $\mathbb{E}(\xi|\mathscr{C})$ when ξ has expectation. Without loss of generality, we may assume $\mathbb{E}\xi^- < \infty$, so that $\mathscr{C} \ni B \mapsto \varphi(B) := \int_B \xi \, \mathrm{d}\mathbb{P}$ is a signed measure with $\varphi \ll \mathbb{P}|_{\mathscr{C}}$. By Theorem 3.51, there exists $\mathbb{P}|_{\mathscr{C}}$-a.s. unique $f \in \mathscr{C}$ such that $\mathrm{d}\varphi = f \, \mathrm{d}\mathbb{P}|_{\mathscr{C}}$, i.e. $\varphi(B) = \int_B f \, \mathrm{d}\mathbb{P} = \int_B \xi \, \mathrm{d}\mathbb{P}$ for $B \in \mathscr{C}$.

Definition 5.4. Let $\mathscr{C} \subset \mathscr{A}$ be a σ-algebra. For any $A \in \mathscr{A}$, $\mathbb{P}(A|\mathscr{C}) = \mathbb{E}(\mathbf{1}_A|\mathscr{C})$ is called the conditional probability of A given $\mathscr{C}$ (with respect to $\mathbb{P}$).

Since the conditional expectation is defined via integral, it inherits most properties of integrals, but in the sense of $\mathbb{P}|_{\mathscr{C}}$-a.s. We collect some of them in the following result, where the convergence theorems can be proved by using the monotone convergence theorem (Exercise 3), as shown in the proofs of the corresponding results for integrals. In the same spirit, some inequalities for integrals and expectations (such as Jensen's inequality, Hölder's inequality, and Minkowski's inequality) also hold for conditional expectations, which are left as exercises as well.

Property 5.5. Assume that the following involved random variables have expectations:

(1) $\mathbb{E}(\mathbb{E}(\xi|\mathscr{C})) = \mathbb{E}\xi$.
(2) If $\xi \in \mathscr{C}$, then $\mathbb{E}(\xi|\mathscr{C}) = \xi$.
(3) (*Monotonicity*) $\xi \leqslant \eta \Rightarrow \mathbb{E}(\xi|\mathscr{C}) \leqslant \mathbb{E}(\eta|\mathscr{C})$.
(4) (*Linear property*) $\mathbb{E}(a\xi + b\eta|\mathscr{C}) = a\mathbb{E}(\xi|\mathscr{C}) + b\mathbb{E}(\eta|\mathscr{C}), a, b \in \mathbb{R}$.
(5) (*Fatou–Lebesgue convergence theorem*) Let η and ζ be integrable. If $\eta \leqslant \xi_n, \mathbb{P}$-a.e. for any $n \geqslant 1$, then $\mathbb{E}\left(\varliminf_{n\to\infty} \xi_n|\mathscr{C}\right) \leqslant \varliminf_{n\to\infty} \mathbb{E}(\xi_n|\mathscr{C})$.
    If $\eta \geqslant \xi_n, \mathbb{P}$-a.e. for any $n \geqslant 1$, then $\varlimsup_{n\to\infty} \mathbb{E}(\xi_n|\mathscr{C}) \leqslant \mathbb{E}\left(\varlimsup_{n\to\infty} \xi_n|\mathscr{C}\right)$.
(6) (*Dominated convergence theorem*) Let η be integrable. If $\eta \leqslant \xi_n \uparrow \xi$ or $|\xi_n| \leqslant \eta$ for any $n \geqslant 1$ and $\xi_n \to \xi$, a.s., then $\mathbb{E}(\xi_n|\mathscr{C}) \to \mathbb{E}(\xi|\mathscr{C})$, a.s.

Property 5.6. Let ξ and η be random variables such that $\eta \in \mathscr{C}$ and $\mathbb{E}\xi\eta$, $\mathbb{E}\xi$ exist. Then $\mathbb{E}(\xi\eta|\mathscr{C}) = \eta\mathbb{E}(\xi|\mathscr{C})$.

Proof. Since both $\mathbb{E}(\xi\eta|\mathscr{C})$ and $\eta\mathbb{E}(\xi|\mathscr{C})$ is $\mathscr{C}$-measurable, by the definition of conditional expectation, we only need to show

$$\int_C \xi\eta \,\mathrm{d}\mathbb{P} = \int_C \eta\mathbb{E}(\xi|\mathscr{C}) \,\mathrm{d}\mathbb{P}, \quad C \in \mathscr{C}.$$

By Theorem 2.12 and Property 5.5(4) and (6), it suffices to prove for $\xi = \mathbf{1}_A, \eta = \mathbf{1}_B, A \in \mathscr{A}, B \in \mathscr{C}$, then the proof is finished since in this case, we have

$$\int_C \eta\mathbb{E}(\xi|\mathscr{C}) \,\mathrm{d}\mathbb{P} = \int_{C\cap B} \xi \,\mathrm{d}\mathbb{P} = \mathbb{P}(A \cap B \cap C) = \int_C \xi\eta \,\mathrm{d}\mathbb{P}. \qquad \square$$

Property 5.7. Let $r \in [1, \infty)$. If $\xi_n \xrightarrow{L^r(\mathbb{P})} \xi$, then $\mathbb{E}(\xi_n|\mathscr{C}) \xrightarrow{L^r(\mathbb{P})} \mathbb{E}(\xi|\mathscr{C})$.

Proof. By Jensen's inequality and the property of conditional expectation, it follows that

$$\begin{aligned}\mathbb{E}|\mathbb{E}(\xi_n|\mathscr{C}) - \mathbb{E}(\xi|\mathscr{C})|^r &= \mathbb{E}|\mathbb{E}(\xi_n - \xi|\mathscr{C})|^r \leqslant \mathbb{E}(\mathbb{E}(|\xi_n - \xi|^r|\mathscr{C})) \\ &= \mathbb{E}|\xi_n - \xi|^r \to 0 \ (n \to \infty). \end{aligned} \qquad \square$$

The following result shows that the conditional expectation of ξ under $\mathscr{C}$ can be regarded as the average of ξ on each atom of $\mathscr{C}$. This property is called smoothness of conditional expectation.

Property 5.8. $\mathbb{E}(\xi|\mathscr{C})$ takes constant value on each atom of $\mathscr{C}$. If $\mathbb{P}(B) > 0$ and B is an atom, then $\mathbb{E}(\xi|\mathscr{C})(\omega) = \frac{1}{\mathbb{P}(B)} \int_B \xi \,\mathrm{d}\mathbb{P}$ for any $\omega \in B$.

Proof. Let B be an atom of $\mathscr{C}$. If $\exists \omega_1, \omega_2 \in B$ such that $\mathbb{E}(\xi|\mathscr{C})(\omega_1) \neq \mathbb{E}(\xi|\mathscr{C})(\omega_2)$, then $\mathscr{C} \ni \{\omega \in B : \mathbb{E}(\xi|\mathscr{C})(\omega) = \mathbb{E}(\xi|\mathscr{C})(\omega_1)\} \subsetneq B$ is nonempty. This contradicts the fact that B is an atom.

Let B be an atom with $\mathbb{P}(B) > 0$. Since $\mathbb{E}(\xi|\mathscr{C})$ takes constant value on B, we have

$$\mathbb{E}(\xi|\mathscr{C})|_B \mathbb{P}(B) = \int_B \mathbb{E}(\xi|\mathscr{C}) \,\mathrm{d}\mathbb{P} = \int_B \xi \,\mathrm{d}\mathbb{P}.$$

Hence, $\mathbb{E}(\xi|\mathscr{C})|_B = \frac{1}{\mathbb{P}(B)} \int_B \xi \,\mathrm{d}\mathbb{P}$. $\square$

The following result shows that the general definition of conditional expectation is consistent with that for $\mathscr{C}$ induced by countable many atoms.

Property 5.9. Let $\{B_n\}_{n\geqslant 1} \subset \mathscr{A}$ be a partition of Ω and $\mathscr{C} = \sigma(\{B_n : n \geqslant 1\})$. Then $\mathbb{E}(\xi|\mathscr{C}) = \sum_{n=1}^{\infty} \mathbb{E}(\xi|B_n)\mathbf{1}_{B_n}$. In particular, $\mathbb{E}(\xi|\mathscr{C}) = \mathbb{E}\xi$ for $\mathscr{C} = \{\phi, \Omega\}$.

Property 5.10. If $\mathscr{C}$ and $\sigma(\xi)$ are independent, then $\mathbb{E}(\xi|\mathscr{C}) = \mathbb{E}\xi$; if $\mathscr{C} \subset \mathscr{C}'$, then

$$\mathbb{E}(\xi|\mathscr{C}) = \mathbb{E}(\mathbb{E}(\xi|\mathscr{C}')|\mathscr{C}).$$

Proof. $\forall B \in \mathscr{C}$, $\mathbf{1}_B$ and ξ are independent, so

$$\int_B \mathbb{E}(\xi|\mathscr{C})\, \mathrm{d}\mathbb{P} = \int_B \xi\, \mathrm{d}\mathbb{P} = \mathbb{E}\mathbf{1}_B\xi = (\mathbb{E}\mathbf{1}_B)\mathbb{E}\xi = \mathbb{P}(B)\mathbb{E}\xi = \int_B \mathbb{E}\xi\, \mathrm{d}\mathbb{P}.$$

Since $B \in \mathscr{C}$ is arbitrary, it follows $\mathbb{E}(\xi|\mathscr{C}) = \mathbb{E}\xi$.

Let $\mathscr{C} \subset \mathscr{C}'$. Then $\forall B \in \mathscr{C}$

$$\int_B \mathbb{E}(\xi|\mathscr{C}')\, \mathrm{d}\mathbb{P} = \mathbb{E}[1_B\mathbb{E}(\xi|\mathscr{C}')] = \mathbb{E}(\mathbb{E}(\xi 1_B|\mathscr{C}')) = \mathbb{E}\xi 1_B = \int_B \xi\, \mathrm{d}\mathbb{P}.$$

Hence, $\mathbb{E}(\xi|\mathscr{C}) = \mathbb{E}(\mathbb{E}(\xi|\mathscr{C}')|\mathscr{C})$. □

Finally, we prove that $\mathbb{E}(\xi|\mathscr{C})$ is the L^2 optimal approximation of ξ among $\mathscr{C}$-measurable functions.

Property 5.11 (Optimal mean square approximation). Let $\xi \in L^2(\mathbb{P}), \mathscr{C} \subset \mathscr{A}$ be a sub-σ-algebra. Then $\mathbb{E}(\xi|\mathscr{C}) \in L^2(\mathbb{P}_{\mathscr{C}})$, and $\mathbb{E}(\xi|\mathscr{C})$ is the optimal approximation of ξ in $L^2(\mathbb{P}_{\mathscr{C}})$: $\forall \eta \in L^2(\mathbb{P}_{\mathscr{C}})$,

$$\mathbb{E}|\xi - \mathbb{E}(\xi|\mathscr{C})|^2 \leqslant \mathbb{E}|\xi - \eta|^2, \quad \mathbb{E}(|\xi - \mathbb{E}(\xi|\mathscr{C})|^2|\mathscr{C}) \leqslant \mathbb{E}(|\xi - \eta|^2|\mathscr{C}),$$

and the equalities hold if and only if $\eta = \mathbb{E}(\xi|\mathscr{C})$, $\mathbb{P}$-a.s.

Proof. We only prove the latter. By Jensen's inequality,

$$|\mathbb{E}(\xi|\mathscr{C})|^2 \leqslant \mathbb{E}(|\xi|^2|\mathscr{C}),$$

so $\mathbb{E}(\xi|\mathscr{C}) \in L^2(\mathbb{P}_{\mathscr{C}})$. $\forall \eta \in L^2(\mathbb{P}_{\mathscr{C}})$, we have

$$\begin{aligned}\mathbb{E}(|\xi-\eta|^2|\mathscr{C}) &= \mathbb{E}(|\xi-\mathbb{E}(\xi|\mathscr{C})|^2|\mathscr{C}) + \mathbb{E}(|\eta-\mathbb{E}(\xi|\mathscr{C})|^2|\mathscr{C})\\ &\quad - 2\mathbb{E}((\eta-\mathbb{E}(\xi|\mathscr{C}))(\xi-\mathbb{E}(\xi|\mathscr{C}))|\mathscr{C}).\end{aligned}$$

Since $\eta - \mathbb{E}(\xi|\mathscr{C}) \in \mathscr{C}$, we have

$$\mathbb{E}((\eta-\mathbb{E}(\xi|\mathscr{C}))(\xi-\mathbb{E}(\xi|\mathscr{C}))|\mathscr{C}) = (\eta-\mathbb{E}(\xi|\mathscr{C}))\mathbb{E}((\xi-\mathbb{E}(\xi|\mathscr{C}))|\mathscr{C}) = 0.$$

Hence, $\mathbb{E}(|\xi-\eta|^2|\mathscr{C}) \geqslant \mathbb{E}(|\xi-\mathbb{E}(\xi|\mathscr{C})|^2|\mathscr{C})$, and the equality holds if and only if $\eta = \mathbb{E}(\xi|\mathscr{C})$, $\mathbb{P}$-a.s. □

5.2 Conditional Expectation Given Function

In this section, we study conditional expectations given the σ-algebra $\mathscr{C} = \sigma(f) = f^{-1}(\mathscr{E})$ induced by a measurable map $f : (\Omega, \mathscr{A}) \to (E, \mathscr{E})$, where $(E, \mathscr{E})$ is a measurable space. In this case, we simply denote $\mathbb{E}(\cdot|\sigma(f))$ by $\mathbb{E}(\cdot|f)$. In particular, when f is a random variable, i.e. $(E, \mathscr{E}) = (\mathbb{R}^n, \mathscr{B}^n)$ for some $n \geqslant 1$, the conditional expectations reflect the influence of f to random variables under study.

Theorem 5.12. *Let ξ be a random variable having expectation, and let $f : (\Omega, \mathscr{A}) \to (E, \mathscr{E})$ be measurable. Then $\mathbb{E}(\xi|f) := \mathbb{E}(\xi|\sigma(f)) = g \circ f$, where $g : E \to \mathbb{R}$ is a measurable function such that*

$$\int_B g \, \mathrm{d}(\mathbb{P} \circ f^{-1}) = \int_{f^{-1}(B)} \xi \, \mathrm{d}\mathbb{P}, \quad B \in \mathscr{E}.$$

Proof. Since $\mathbb{E}(\xi|\sigma(f))$ is $\sigma(f)$-measurable, by Theorem 2.22, there exists a measurable function $g : (E, \mathscr{E}) \to (\mathbb{R}, \mathscr{B})$ such that $\mathbb{E}(\xi|\sigma(f)) = g \circ f$. Combining this with the integral transform formula (Theorem 3.27) and the definition of conditional expectation, we derive the desired formula. □

By taking $\xi = \mathbf{1}_A (A \in \mathscr{A})$ in Theorem 5.12, we obtain the following result.

Corollary 5.13. *For any $A \in \mathscr{A}$, we have $\mathbb{P}(A|\sigma(f)) = g \circ f$, where $g : (E, \mathscr{E}) \to (\mathbb{R}, \mathscr{B})$ is measurable, satisfying*

$$\int_B g \, \mathrm{d}(\mathbb{P} \circ f^{-1}) = \mathbb{P}(A \cap f^{-1}(B)), \quad \forall B \in \mathscr{E}.$$

As explained in the beginning of this chapter, the conditional expectation given an event B can be formulated as the integral with respect to the conditional probability $\mathbb{P}(\cdot|B)$. As we have already defined the conditional expectation $\mathbb{E}(\cdot|\mathscr{C})$ and the conditional probability $\mathbb{P}(\cdot|\mathscr{C})$, we wish to establish the same link of them, i.e. to formulate $\mathbb{E}(\xi|\mathscr{C})$ as the integral of ξ with respect to $\mathbb{P}(\cdot|\mathscr{C})$. However, for each event A, $\mathbb{P}(A|\mathscr{C})$ is only $\mathbb{P}$-a.s. defined. So, to establish the desired formula, we need to fix a $\mathbb{P}$-version (i.e. a pointwise defined function) in the class of of $\mathbb{P}(A|\mathscr{C})$, denoted by $\mathbb{P}^{\mathscr{C}}(\cdot, A)$. If $\mathbb{P}^{\mathscr{C}}$ happens to be a transition probability on $(\Omega, \mathscr{C}) \times (\Omega, \mathscr{A})$, then we will be able to verify the formula $\mathbb{E}(\xi|\mathscr{C}) = \int_\Omega \xi \, \mathrm{d}\mathbb{P}^{\mathscr{C}}$. Such a transition probability is called the regular conditional probability given $\mathscr{C}$.

5.3 Regular Conditional Probability

5.3.1 Definition and properties

Definition 5.14 (Regular conditional probability). Let $\mathscr{C} \subset \mathscr{A}$ be a sub-σ-algebra of $\mathscr{A}$. A transition probability $\mathbb{P}^{\mathscr{C}}$ on $(\Omega, \mathscr{C}) \times (\Omega, \mathscr{A})$ is called the regular conditional probability given $\mathscr{C}$ (with respect to $\mathbb{P}$) if

$$\mathbb{P}^{\mathscr{C}}(\cdot, A) = \mathbb{E}(\mathbf{1}_A|\mathscr{C}) = \mathbb{P}(A|\mathscr{C}), \quad \forall A \in \mathscr{A}.$$

Obviously, the regular conditional expectation is $\mathbb{P}$-a.s. unique. If it exists, we may formulate the conditional expectation given $\mathscr{C}$ by the integral with respect to $\mathbb{P}^{\mathscr{C}}$.

Theorem 5.15. *Let $\mathbb{P}^{\mathscr{C}}$ be the regular conditional probability given $\mathscr{C}$. Then for any random variable ξ having expectation,*

$$\mathbb{E}(\xi|\mathscr{C}) = \int_\Omega \xi \mathbb{P}^{\mathscr{C}}(\cdot, \mathrm{d}\omega).$$

Proof. By definition, formula holds for ξ being an indicator function. Then the proof is finished by Theorem 2.12, Theorem 3.5 and the linearity of integrals. □

By the link of the regular conditional probability and the conditional expectation, properties of conditional expectation can be formulated by using the regular conditional probability. In the following, we only reformulate one property as example.

Theorem 5.16. *Let $\mathscr{C} \subset \mathscr{C}' \subset \mathscr{A}$ be a sub-σ-algebras, and let $\mathbb{P}^{\mathscr{C}}$ and $\mathbb{P}^{\mathscr{C}'}$ be the associated regular conditional probabilities. Then for any random variable $\xi \in \mathscr{C}$ and $\xi' \in \mathscr{C}'$ such that $\mathbb{E}\xi\xi'$ and $\mathbb{E}\xi$ exist, there holds*

$$\int (\xi'\xi)(\omega)\mathbb{P}^{\mathscr{C}}(\cdot, \mathrm{d}\omega) = \int \xi'(\omega') \left[\int \xi(\omega)\mathbb{P}^{\mathscr{C}'}(\omega', \mathrm{d}\omega) \right] \mathbb{P}^{\mathscr{C}}(\cdot, \mathrm{d}\omega').$$

By the integral transformation theorem, the expectation of a random variable ξ can be formulated as the integral of identity function with respect to the distribution of ξ, which only depends on the restricted probability measure on the σ-algebra induced by ξ. Correspondingly, in the following, we introduce the regular conditional distribution and mixed conditional distribution of ξ given $\mathscr{C}$.

5.3.2 Conditional distribution

Definition 5.17. Let $\xi_T = \{\xi_t : t \in T\}$ be a family of random variables on $(\Omega, \mathscr{A}, \mathbb{P})$, and let $\mathscr{C} \subset \mathscr{A}$ be a sub-σ-algebra of $\mathscr{A}$.

(1) A transition probability on $(\Omega, \mathscr{C}) \times (\Omega, \sigma(\xi_T))$ is called the regular conditional distribution of ξ_T under $\mathscr{C}$ if

$$\mathbb{P}^{\mathscr{C}}(\cdot, A) = \mathbb{P}(A|\mathscr{C}), \quad A \in \sigma(\xi_T).$$

(2) A transition probability $\mathbb{P}^{\mathscr{C}}_{\xi_T}$ on $(\Omega, \mathscr{C}) \times (\mathbb{R}^T, \mathscr{B}^T)$ is called a mixed conditional distribution of ξ_T under $\mathscr{C}$ if

$$\mathbb{P}^{\mathscr{C}}_{\xi_T}(\cdot, B) = \mathbb{P}(\xi_T^{-1}(B)|\mathscr{C}), \quad B \in \mathscr{B}^T.$$

Theorem 5.18. *Let $g : \mathbb{R}^T \to \mathbb{R}$ be Borel measurable such that $\mathbb{E}g(\xi_T)$ exists. Let $\mathbb{P}^{\mathscr{C}}$ and $\mathbb{P}^{\mathscr{C}}_{\xi_T}$ be, respectively, the regular conditional* *distribution and mixed conditional distribution of ξ given $\mathscr{C}$ exist, then*

$$\mathbb{E}(g(\xi_T)|\mathscr{C}) = \int_\Omega g(\xi_T(\omega))\mathbb{P}^{\mathscr{C}}(\cdot, \mathrm{d}\omega) = \int_{\mathbb{R}^T} g(x_T)\mathbb{P}^{\mathscr{C}}_{\xi_T}(\cdot, \mathrm{d}x_T).$$

Proof. As explained many times, the desired formulae follow from those with $g = \mathbf{1}_B, B \in \mathscr{B}^T$. □

Theorem 5.19. *If the conditional distribution of ξ_T given $\mathscr{C}$ exists, then its mixed conditional distribution exists too. When $\xi_T(\Omega) \in \mathscr{B}^T$, the converse assertion also holds.*

Proof. Let $\mathbb{P}^{\mathscr{C}}$ be the conditional distribution of ξ_T given $\mathscr{C}$. Then

$$\mathbb{P}^{\mathscr{C}}_{\xi_T}(\omega, B) = \mathbb{P}^{\mathscr{C}}(\omega, \xi_T^{-1}(B)), \quad B \in \mathscr{B}^T$$

gives the mixed conditional distribution. Conversely, if $\mathbb{P}^{\mathscr{C}}_{\xi_T}$ exists and $\xi_T(\Omega) \in \mathscr{B}^T$, then for any $A \in \sigma(\xi_T)$, there exists $B \in \mathscr{B}^T$ such that $A = \xi_T^{-1}(B)$, so that $\xi_T(A) = B \cap \xi_T(\Omega) \in \mathscr{B}^T$. Hence, we can define the regular conditional distribution as $\mathbb{P}^{\mathscr{C}}(\cdot, A) = \mathbb{P}^{\mathscr{C}}_{\xi_T}(\cdot, \xi_T(A))$. □

5.3.3 Existence of regular conditional probability

We first prove the existence of mixed conditional distribution.

Theorem 5.20. *Let $\xi = (\xi_1, \xi_2, \ldots, \xi_n)$ be an n-dimensional random variable on $(\Omega, \mathscr{A}, \mathbb{P})$ and $\mathscr{C}$ be a sub-σ-algebra of $\mathscr{A}$. Then $\mathbb{P}^{\mathscr{C}}_\xi$ exists, hence when $\xi(\Omega) \in \mathscr{B}^n$, the conditional distribution of ξ under $\mathscr{C}$ exists.*

Proof. To construct $\mathbb{P}^{\mathscr{C}}_\xi(\omega, \cdot)$, we only need to determine the corresponding probability distribution function. By the left continuity, it suffices to fix the distribution function on a countable dense subset of $\bar{\mathbb{R}}^n$, say, on the rational number space $\mathbb{Q}^n$. For any $r \in (\mathbb{Q} \cup \{\infty\})^n$, we fix a $\mathscr{C}$-measurable function $F(\cdot; r)$ in the class $\mathbb{P}(\xi < r|\mathscr{C})$. Obviously, F has the following properties: there exists a $\mathbb{P}$-null set N such that

(1) $\forall a, b \in \mathbb{Q}^n, a \leqslant b$, $\Delta_{b,a}F(\omega; \cdot) = \mathbb{P}(\xi \in [a, b)|\mathscr{C})(\omega) \geqslant 0$ and $F(\omega; b) \geqslant F(\omega; a), \quad \omega \notin N$;
(2) $\lim_{\mathbb{N} \ni m \to \infty} F(\omega; m, \ldots, m) = 1, \quad \omega \notin N$;

(3) for any $1 \leqslant i \leqslant n$, let $r_m^{(i)} \in \bar{\mathbb{Q}}^n$ such that the ith component is $-m$ and others are ∞; then $\lim_{\mathbb{N}\ni m\to\infty} F(\omega; r_m) = 0, \quad \omega \notin N$;

(4) $\forall r_0 \in \mathbb{Q}^n, \lim_{\mathbb{N}\ni m\to\infty} F(\omega; r_0 - \frac{1}{m}) = F(\omega; r_0), \quad \omega \notin N.$

Let

$$F^{\mathscr{C}}(\omega; r) = \begin{cases} F(\omega; r), & \omega \in N^c; \\ \mathbf{1}_{(0,\infty)}(r), & \omega \in N, r \in \mathbb{Q}^n. \end{cases}$$

Moreover, for each $x \in \mathbb{R}^n$, let $F^{\mathscr{C}}(\omega; x) = \lim_{r\uparrow x} F^{\mathscr{C}}(\omega; r)$, which is well defined by the increasing property. Then $\forall \omega \in \Omega, F^{\mathscr{C}}(\omega; \cdot)$ is a probability distribution function, so it induces a unique probability measure $\mathbb{P}_\xi^{\mathscr{C}}(\omega; \cdot)$ on $(\mathbb{R}^n, \mathscr{B}^n)$ such that $\mathbb{P}_\xi^{\mathscr{C}}(\omega; (-\infty, x)) = F^{\mathscr{C}}(\omega; x)$ for $\omega \in \Omega$ and $x \in \mathbb{R}^n$.

Finally,

$$\Pi = \{(-\infty, r) : r \in \mathbb{Q}^n\}, \ \Lambda = \left\{B \in \mathscr{B}^n : \mathbb{P}_\xi^{\mathscr{C}}(\cdot, B) = \mathbb{P}(\xi \in B|\mathscr{C})\right\}.$$

Then Π is a π-system, $\Lambda \supset \Pi$, and Λ is a λ-system. By the monotone class theorem, we obtain $\Lambda = \mathscr{B}^n$, so that $\mathbb{P}_\xi^{\mathscr{C}}$ is the mixed conditional distribution of ξ given $\mathscr{C}$. □

As a consequence of Theorem 5.20, we confirm the existence of regular probability measure for $(\Omega, \mathscr{A}, \mathbb{P}) = (\mathbb{R}^n, \mathscr{B}^n, \mathbb{P})$.

Theorem 5.21. *Let $(\Omega, \mathscr{A}, \mathbb{P}) = (\mathbb{R}^n, \mathscr{B}^n, \mathbb{P})$. Then for any sub-$\sigma$-algebra $\mathscr{C}$ of $\mathscr{B}^n$, there exists the regular conditional probability $\mathbb{P}^{\mathscr{C}}$.*

Proof. Let $\xi(x) = x$ for $x \in \mathbb{R}^n$. Then $\sigma(\xi) = \mathscr{B}^n$ and $\mathbb{P}_\xi^{\mathscr{C}}$ is just the regular conditional probability of $\mathbb{P}$ given $\mathscr{C}$. □

As an application of the regular conditional probability, any probability measure on $\mathbb{R}^n$ can be induced by a marginal distribution together with some transition probabilities as in Theorem 4.12.

Theorem 5.22. *Let $\mathbb{P}$ be a probability on $(\mathbb{R}^n, \mathscr{B}^n)$. Then there exist probability $\mathbb{P}_1$ on $\mathscr{B}$ and transition probabilities $\mathbb{P}_k(x_1, x_2, \ldots, \mathrm{d}x_k)$ on $\mathbb{R}^{k-1} \times \mathscr{B}$ for $k = 2, \ldots, n$ such that*

$$\mathbb{P}(B) = \int_{\mathbb{R}} \cdots \int_{\mathbb{R}} \mathbf{1}_B(x_1, \ldots, x_n) \mathbb{P}_n(x_1, \ldots, x_{n-1}, \mathrm{d}x_n) \cdots \mathbb{P}_1(\mathrm{d}x_1), B \in \mathbb{R}^n.$$

Proof. By induction, we only prove for $n = 2$. In this case, let $\mathscr{C} = \{A \times \mathbb{R} : A \in \mathscr{B}\}$, and for any $B_1, B_2 \in \mathscr{B}$ and $B \in \mathscr{B}^2$,

$$\mathbb{P}_1(B_1) = \mathbb{P}(B_1 \times \mathbb{R}), \quad \mathbb{P}_2(x_1, B_2) = \mathbb{P}^{\mathscr{C}}((x_1, 0), \mathbb{R} \times B_2),$$
$$\tilde{\mathbb{P}}(B) = \int_{\mathbb{R}} \mathrm{d}\mathbb{P}_1(x_1) \int_{\mathbb{R}} \mathbf{1}_B \mathbb{P}_2(x_1, \mathrm{d}x_2).$$

Noting that $\mathbb{P}_2(x_1, B_2) = \mathbb{P}^{\mathscr{C}}((x_1, x_2), \mathbb{R} \times B_2)$ for any $x_2 \in \mathbb{R}$ since $\{(x_1, x_2) : x_1 \in \mathbb{R}\}$ is an atom of $\mathscr{C}$ on which the conditional probability is constant, we obtain

$$\begin{aligned}\mathbb{P}(B_1 \times B_1) &= \int_{\mathbb{R}} \mathbf{1}_{B_1}(x_1)\mathbb{P}_1(\mathrm{d}x_1) \int_{\mathbb{R}} \mathbf{1}_{B_2}(x_2)\mathbb{P}_2(x_1, \mathrm{d}x_2)\\ &= \int_{\mathbb{R}^2} \mathbf{1}_{B_1 \times \mathbb{R}}(x_1, x_2)\mathbb{P}_2(x_1, B_2)\mathbb{P}(\mathrm{d}x_1, \mathrm{d}x_2)\\ &= \mathbb{E}[\mathbb{E}(1_{B_1 \times B_2}|\mathscr{C})] = \mathbb{E}1_{B_1 \times B_2} = \mathbb{P}(B_1 \times B_2).\end{aligned}$$

This finishes the proof by the uniqueness in the measure extension theorem. □

5.4 Kolmogorov's Consistent Theorem

In this section, we construct probability measures on an infinite product space by using a family of consistent probability measures on finite product spaces. For this, we introduce the concept of consistency. Let T be an infinite index set, and let $(\Omega_t, \mathscr{A}_t)$ be a measurable space for every $t \in T$. Recall that $\forall T' \subset T$, $\Omega^{T'} := \prod_{t \in T'} \Omega_t$, $\mathscr{A}^{T'} := \prod_{t \in T'} \mathscr{A}_t$. If S is a finite subset of T, we write $S \Subset T$.

Definition 5.23. The family of probability measures $\{\mathbb{P}^S : S \Subset T\}$ is called consistent if each $\mathbb{P}^S$ is a probability measure on $\mathscr{A}^S$ and

$$\mathbb{P}^S(A^S) = \mathbb{P}^{S'}\left(A^S \times \Omega^{S'-S}\right), \quad A^S \in \mathscr{A}^S, \quad S \subset S' \Subset T.$$

Theorem 5.24 (Kolmogorov's consistent theorem). *Let* $(\Omega_t, \mathscr{A}_t) = (\mathbb{R}, \mathscr{B})$ *for* $t \in T$, *and let* $\{\mathbb{P}^S : S \Subset T\}$ *be a family of consistent probability measures. Then there exists a unique probability measure* $\mathbb{P}$ *on* $(\mathbb{R}^T, \mathscr{B}^T)$ *such that*

$$\mathbb{P}(B^S \times \mathbb{R}^{T-S}) = \mathbb{P}^S(B^S), \quad S \Subset T, \quad B^S \in \mathscr{B}^S.$$

Proof. (1) By the consistency, it is easy to see that $\mathbb{P}$ is a well-defined finite-additive measure on the class $\mathscr{C}^T$ of measurable cylindrical sets and $\mathbb{P}(\mathbb{R}^T) = 1$. So, by the measure extension theorem, it suffices to verify the σ-additivity of $\mathbb{P}$.

(2) When T is countable, we may let $T = \mathbb{N}$. By Theorem 5.22, there exist probability $\mathbb{P}_1$ on $\mathbb{R}$ and transition probabilities $\mathbb{P}_n$ on $\mathbb{R}^{n-1} \times \mathscr{B}$ for each $n \geqslant 2$ such that $\mathbb{P}^{\{1,2,\ldots,n\}} = \mathbb{P}_n \cdot \mathbb{P}_{n-1} \cdots \mathbb{P}_1$, $\forall n \geqslant 2$. Then the desired assertion follows from Tulcea's theorem.

(3) In general, we only need to prove that $\mathbb{P}$ is continuous at $\varnothing$. Let $\{A_n\}_{n \geqslant 1} \subset \mathscr{C}^T$ with $A_n \downarrow \varnothing$. For any $n \geqslant 1$, there exists $T_n \Subset T$ such that $A_n = A^{T_n} \times \mathbb{R}^{T \setminus T_n}$ and $A^{T_n} \in \mathscr{B}^{T_n}$. Set $T_\infty = \bigcup_{n=1}^{\infty} T_n$. Then T_∞ is countable. By step (2), $\mathbb{P}$ is σ-additive on the algebra $\mathscr{C}^{T_\infty} \times \mathbb{R}^{T \setminus T_\infty}$, so that Theorem 1.34 implies that $\mathbb{P}(A_n) \downarrow 0$ $(n \to \infty)$. □

Remark 5.25. The proof of Theorem 5.24 mainly uses Theorem 5.22 and Tulcea's theorem, where the latter works for general $(\Omega_t, \mathscr{A}_t)$. Note that Theorem 5.22 can be extended to a Polish space in place of $\mathbb{R}^n$, see Exercise 20 in this chapter. So, Theorem 5.24 can be extended to the case that each $(\Omega_t, \mathscr{A}_t)$ is a Polish space.

5.5 Exercises

1. Prove Proposition 5.2.
2. For a sequence of random variables $0 \leqslant \xi_n \uparrow \xi$, prove $\mathbb{E}(\xi_n|\mathscr{C}) \uparrow \mathbb{E}(\xi|\mathscr{C})$.
3. Prove Property 5.5.
4. Prove Hölder's inequality $\mathbb{E}(\xi\eta|\mathscr{C}) \leqslant \mathbb{E}(|\xi|^p|\mathscr{C})^{1/p}\mathbb{E}(|\eta|^q|\mathscr{C})^{1/q}$, $p > 1, \frac{1}{p} + \frac{1}{q} = 1$.

5. Formulate and prove Jensen's and Minkowski's inequalities for conditional expectations.

6. Prove Corollary 5.9.

7. Construct a probability space $(\Omega, \mathscr{A}, \mathbb{P})$, sub-$\sigma$-algebras $\mathscr{C}_1$ and $\mathscr{C}_2$ of $\mathscr{A}$, and an integrable random variable ξ such that $\mathbb{E}(\xi|\mathscr{C}_1 \cap \mathscr{C}_2) \neq \mathbb{E}(\mathbb{E}(\xi|\mathscr{C}_1)|\mathscr{C}_2)$.

8. Denote total rational numbers by $x_1, x_2, \ldots$ and let
$$F(x) = \sum_{n=1}^{\infty} 2^{-n} \mathbf{1}_{(x_n,\infty)}(x), \quad x \in \mathbb{R}.$$
Prove that F is a probability distribution function on $\mathbb{R}$.

9. (Martingale) Let $\{\mathscr{A}_n\}_{n\geqslant 1}$ be a sequence of increasing sub-σ-algebras of $\mathscr{A}$. If a sequence of random variables $\{\xi_n\}_{n\geqslant 1}$ satisfies
$$\mathbb{E}(\xi_{n+1}|\mathscr{A}_n) = \xi_n, \quad n \geqslant 1,$$
then it is called a martingale sequence. For an integrable random variable ξ, prove that $\xi_n = \mathbb{E}(\xi|\mathscr{A}_n)$ is a martingale sequence.

10. (Markov chain) Let $\{\xi_n\}_{n\geqslant 1}$ be a sequence of random variables. Set $\{\mathscr{A}_n = \sigma(\{\xi_m : m \leqslant n\})\}$. If
$$\mathbb{E}(\xi_{n+1}|\mathscr{A}_n) = \mathbb{E}(\xi_{n+1}|\xi_n), \quad n \geqslant 1,$$
then $\{\xi_n\}_{n\geqslant 1}$ is called a Markov chain. Let $\{X_n\}_{n\geqslant 1}$ be a sequence of independent random variables. Prove that $\{\xi_n = \sum_{m=1}^{n} X_m\}$ is a Markov chain.

11. Let $\{\xi_n\}_{n\geqslant 1}$ be a sequence of random variables, and let
$$\mathscr{A}_n = \sigma(\{\xi_m : m \leqslant n\}), \quad \mathscr{A}^n = \sigma(\{\xi_m : m \geqslant n\}), \quad n \geqslant 1.$$
Prove that $\{\xi_n\}_{n\geqslant 1}$ is a Markov chain if and only if one of the following conditions holds:

(a) $\mathbb{E}(\xi_m|\mathscr{A}_n) = \mathbb{E}(\xi_m|\xi_n), \quad m \geqslant n \geqslant 1;$
(b) $\mathbb{E}(\eta|\mathscr{A}_n) = \mathbb{E}(\eta|\xi_n), \quad \eta \in \mathscr{A}^n, \quad n \geqslant 1;$

(c) $\forall \eta \in \mathscr{A}_n, \zeta \in \mathscr{A}^n$ such that $\eta, \zeta, \eta\zeta$ are integrable, there holds

$$\mathbb{E}[\eta\zeta|\xi_n] = \mathbb{E}[\eta|\xi_n]\mathbb{E}[\zeta|\xi_n], \quad n \geqslant 1.$$

12. Let matrix $P = (p_{ij})_{i,j=0}^{\infty}$ satisfy $p_{ij} \geqslant 0$, $\sum_{j=0}^{\infty} p_{ij} = 1$. Construct a probability space $(\Omega, \mathscr{A}, \mathbb{P})$ and a sequence of random variables $\{\xi_n\}_{n\geqslant 0}$ such that $\mathbb{P}(\xi_{n+1} = j|\xi_n = i) = p_{ij}$ for $n \geqslant 0$ and $i, j \geqslant 0$.

13. Let ξ and η be random variables such that $\mathbb{E}(\xi|\mathscr{C}) = \eta$ and $\mathbb{E}\xi^2 = \mathbb{E}\eta^2 < \infty$. Prove that $\xi = \eta$, a.s.

14. Let $\xi \in L^1(\mathbb{P})$. Prove that the family of random variables

$$\{\mathbb{E}(\xi|\mathscr{C}) : \mathscr{C} \text{ is sub-}\sigma\text{-algebra of } \mathscr{A}\}$$

is uniformly integrable.

15. Let ξ and η be independent identically distributed such that $\mathbb{E}\xi$ exists. Prove $\mathbb{E}(\xi|\xi + \eta) = (\xi + \eta)/2$.

16. Let $(\Omega, \mathscr{A}, \mathbb{P})$ be a probability space, let $(E, \mathscr{E})$ be a measurable space, and let $T : \Omega \to E$ be a measurable map. Prove that for any sub-σ-algebra $\mathscr{C}$ of $\mathscr{E}$, there holds

$$\mathbb{P}\left(T^{-1}(B)|T^{-1}(\mathscr{C})\right) = (\mathbb{P} \circ T^{-1})(B|\mathscr{C}) \circ T, \quad B \in \mathscr{E}.$$

17. For an event $\mathbb{P}(A) > 0$, denote $\mathbb{P}^A = \mathbb{P}(\cdot|A)$. Prove that for any $B \in \mathscr{A}$ and sub-σ-algebra $\mathscr{C}$ of $\mathscr{A}$,

$$\mathbb{P}^A(B|\mathscr{C}) = \frac{\mathbb{P}(A \cap B|\mathscr{C})}{\mathbb{P}(A|\mathscr{C})}.$$

18. (a) Let $\mathscr{C}_1 \subset \mathscr{C}_2$ be sub-σ-algebras of $\mathscr{A}$, and let ξ be a random variable with $\mathbb{E}\xi^2 < \infty$. Prove

$$\mathbb{E}((\xi - \mathbb{E}(\xi|\mathscr{C}_1))^2) \geqslant \mathbb{E}((\xi - \mathbb{E}(\xi|\mathscr{C}_2))^2).$$

(b) Let $\mathrm{Var}(\xi|\mathscr{C}) = \mathbb{E}((\xi - \mathbb{E}(\xi|\mathscr{C}))^2|\mathscr{C})$. Prove

$$\mathrm{Var}(\xi) = \mathbb{E}(\mathrm{Var}(\xi|\mathscr{C})) + \mathrm{Var}(\mathbb{E}(\xi|\mathscr{C})).$$

19. Let $\mathscr{C}_i, i = 1,2,3$ be sub-σ-algebras of $\mathscr{A}$, and let $\mathscr{C}_{ij} = \sigma(\mathscr{C}_i \cup \mathscr{C}_j), 1 \leqslant i,j \leqslant 3$. Prove that the following statements are equivalent to each other:
    (a) $\mathbb{P}(A_3|\mathscr{C}_{12}) = \mathbb{P}(A_3|\mathscr{C}_2), \quad \forall A_3 \in \mathscr{C}_3;$
    (b) $\mathbb{P}(A_1 \cap A_3|\mathscr{C}_2) = \mathbb{P}(A_1|\mathscr{C}_2)\mathbb{P}(A_3|\mathscr{C}_2), \quad \forall A_1 \in \mathscr{C}_1, \quad A_3 \in \mathscr{C}_3;$
    (c) $\mathbb{P}(A_1|\mathscr{C}_{23}) = \mathbb{P}(A_1|\mathscr{C}_2), \quad \forall A_1 \in \mathscr{C}_1.$

20. Let $\mathbb{P}$ be a probability on a Polish (i.e. complete separable metric) space E. Then for any sub-σ-algebra $\mathscr{C}$ of the Borel σ-algebra $\mathscr{B}(E)$, the regular conditional probability $\mathbb{P}^{\mathscr{C}}$ exists.

Chapter 6

Characteristic Function and Weak Convergence

We have learnt the characteristic function of a random variable, which is determined by the distribution function according to the L–S representation of expectation and has better analysis properties than the distribution function. In this chapter, we study characteristic functions for general finite measures on $\mathbb{R}^n$ and establish an inverse formula to show that the characteristic function of a random variable also determines the distribution function. Therefore, we can use the convergence of characteristic functions to define the convergence of finite measures or random variables, which is called the weak convergence, and the associated topology on the space of finite measures is called weak topology. More generally, we will introduce several different type convergences for finite measures on a metric space and present some equivalent statements for the weak convergence. In particular, the weak convergence for the distributions of random variables is equivalent to the convergence in distribution.

6.1 Characteristic Function of Finite Measure

6.1.1 Definition and properties

Definition 6.1. Let μ be a finite measure on $(\mathbb{R}^n, \mathscr{B}^n)$. The characteristic function (or Fourier–Stieltjes transform) of μ is defined as

$$f_\mu(t) = \int_{\mathbb{R}^n} \mathrm{e}^{\mathrm{i}\langle t,x\rangle} \mu(\mathrm{d}x), \quad t \in \mathbb{R}^n.$$

Obviously, the characteristic function has the following properties.

Property 6.2. Let $\bar{a}$ be the conjugate number of $a \in \mathbb{C}$.

(1) Let μ be a finite measure on $\mathbb{R}^n$. Then for any $t \in \mathbb{R}^n$, we have

$$|f_\mu(t)| \leqslant f_\mu(0) = \mu(\mathbb{R}^n), \quad \bar{f}_\mu(t) = f_\mu(-t)$$

and the increment inequality

$$|f_\mu(t) - f_\mu(t+h)|^2 \leqslant 2f_\mu(0)[f_\mu(0) - \mathrm{Re} f_\mu(h)], \quad h \in \mathbb{R}^n.$$

Consequently, f_μ is uniformly continuous.

(2) Let μ_k be a finite measure on $\mathbb{R}^{m_k}$ $(k = 1, 2, \ldots, n)$, and let $\mu := \prod_{k=1}^n \mu_i$. Then

$$f_\mu(t) = \prod_{k=1}^n f_{\mu_k}\left(t^{(m_k)}\right), \; t = \left(t^{(m_1)}, \ldots, t^{(m_n)}\right) \in \mathbb{R}^{m_1+\cdots+m_n}.$$

Proof. We only prove the the increment inequality, since other assertions are obvious. Since $f_\mu(0) = \mu(\mathbb{R}^n)$, by the Schwarz inequality, we have

$$\begin{aligned}
|f_\mu(t) - f_\mu(t+h)|^2 &\leqslant f_\mu(0) \int_{\mathbb{R}^n} \left|\mathrm{e}^{\mathrm{i}\langle t,x\rangle} - \mathrm{e}^{\mathrm{i}\langle t+h,x\rangle}\right|^2 \mu(\mathrm{d}x) \\
&\leqslant f_\mu(0) \int_{\mathbb{R}^n} \left|\mathrm{e}^{\mathrm{i}\langle h,x\rangle} - 1\right|^2 \mu(\mathrm{d}x) \\
&= 2f_\mu(0) \int_{\mathbb{R}^n} (1 - \cos\langle h, x\rangle)\mu(\mathrm{d}x) \\
&= 2f_\mu(0)(f(0) - \mathrm{Re} f_\mu(h)).
\end{aligned}$$

□

Finally, we characterize the derivatives of f_μ.

Proposition 6.3. *Let μ be a finite measure on $(\mathbb{R}, \mathscr{B})$ and let $n \geqslant 1$. If $\int_{\mathbb{R}} |x|^n \mu(\mathrm{d}x) < \infty$, then f_μ has derivatives up to n-th order, and $\forall 0 \leqslant k \leqslant n$,*

$$f^{(k)}(t) = i^k \int_{\mathbb{R}} x^k e^{itx} \mu(\mathrm{d}x), \quad t \in \mathbb{R}.$$

In particular,

$$\int_{\mathbb{R}} x^k \mu(\mathrm{d}x) = i^{-k} f^{(k)}(0).$$

6.1.2 Inverse formula

In this part, we aim to determine a finite measure by using its characteristic function via the following inverse formula. An interval $[a, b]$ in $\mathbb{R}^n$ is called μ-continuous for a finite measure μ on $\mathbb{R}^n$ if $\mu(\partial[a, b]) = 0$.

Theorem 6.4 (Inverse formula). *Let μ be a finite measure on $(\mathbb{R}^n, \mathscr{B}^n)$. Then for any μ-continuous interval $[a, b]$ in $\mathbb{R}^n$, we have*

$$\mu([a,b]) = \lim_{T\to\infty} \frac{1}{(2\pi)^n} \int_{[-T,T]^n} \prod_{k=1}^n \frac{e^{-it_k a_k} - e^{-it_k b_k}}{it_k} f(t_1, \ldots, t_n) \mathrm{d}t_1 \cdots \mathrm{d}t_n.$$

Proof. Let $I(T)$ denote the integral in the right-hand side over $[-T, T]^n$. By the definition of f_μ and Fubini's theorem, we obtain

$$\begin{aligned}
I(T) &= \int_{\mathbb{R}^n} \mu(\mathrm{d}x) \int_{-T}^{T} \cdots \int_{-T}^{T} \prod_{k=1}^n \frac{\mathrm{e}^{-\mathrm{i}t_k a_k} - \mathrm{e}^{-\mathrm{i}t_k b_k}}{\mathrm{i}t_k} \mathrm{e}^{\mathrm{i}\sum_{k=1}^n t_k x_k} \mathrm{d}t_1 \cdots \mathrm{d}t_n \\
&= \int_{\mathbb{R}^n} \left(\prod_{k=1}^n \int_{-T}^{T} \frac{\mathrm{e}^{-\mathrm{i}t_k a_k} - \mathrm{e}^{-\mathrm{i}t_k b_k}}{\mathrm{i}t_k} \mathrm{e}^{\mathrm{i}t_k x_k} \mathrm{d}t_k \right) \mu(\mathrm{d}x) \\
&= 2^n \int_{\mathbb{R}^n} \left(\prod_{k=1}^n \int_0^T \frac{\sin t_k(x_k - a_k) - \sin t_k(x_k - b_k)}{t_k} \mathrm{d}t_k \right) \mu(\mathrm{d}x) \\
&= 2^n \int_{\mathbb{R}^n} \left(\prod_{k=1}^n \int_{T(x_k - b_k)}^{T(x_k - a_k)} \frac{\sin t}{t} \mathrm{d}t \right) \mu(\mathrm{d}x).
\end{aligned}$$

Since $\int_s^r \frac{\sin t}{t} \mathrm{d}t$ is bounded in $s \leqslant r \in \mathbb{R}$, and $\int_{-\infty}^{\infty} \frac{\sin t}{t} \mathrm{d}t = \pi$, the dominated convergence theorem implies

$$\lim_{T\to\infty} I(T) = (2\pi)^n \mu((a, b)) = (2\pi)^n \mu([a, b]). \qquad \square$$

To prove that μ is uniquely determined by f_μ via Theorem 6.4, we need to show that μ has plentiful enough continuous intervals, or only rare intervals are not μ-continuous. To this end, we present the following result.

Lemma 6.5. *Let μ be a finite measure on $\mathbb{R}^n$. Then the set*

$$D(\mu) := \{a \in \mathbb{R} : \exists k \in \{1, \ldots, n\} \text{ such that } \mu(\{x : x_k = a\}) > 0\}$$

is at most countable.

Proof. Let

$$D_{m,k}(\mu) = \left\{a \in \mathbb{R} : \mu(\{x : x_k = a\}) \geqslant \frac{1}{m}\right\}, \quad m \geqslant 1, 1 \leqslant k \leqslant n.$$

Then $D(\mu) = \bigcup_{k,m} D_{m,k}(\mu)$. As μ is finite, each $D_{m,k}(\mu)$ is a finite set, so that $D(\mu)$ is at most countable. □

Lemma 6.6. *If an interval $[a, b]$ in $\mathbb{R}^n$ is such that all components of a and b are in the set $C(\mu) := \mathbb{R} \backslash D(\mu)$, then it is μ-continuous.*

Proof. Let $a = (a_k)_{1 \leqslant k \leqslant n}$ and $b = (b_k)_{1 \leqslant k \leqslant n}$ such that $\{a_k, b_k : 1 \leqslant k \leqslant n\} \subset C(\mu)$. Then

$$\partial[a, b] \subset \bigcup_{k=1}^{n} \{x_k = a_k \text{ or } b_k\}$$

is a μ-null set. □

Proposition 6.7. *Let μ_1 and μ_2 be finite measures on $\mathbb{R}^n$. If μ_1 and μ_2 are equal on their common continuous intervals, then $\mu_1 = \mu_2$. Consequently, a finite measure on $\mathbb{R}^n$ is uniquely determined by its characteristic function.*

Proof. By Theorem 6.4, we only need to prove the first assertion. As $D(\mu_1) \cup D(\mu_2)$ is at most countable, $C := C(\mu_1) \cap C(\mu_2)$ is dense in $\mathbb{R}$. $\forall [a, b) \subset \mathbb{R}^n$ and $1 \leqslant k \leqslant n, \exists \{a_k^{(m)}\}, \{b_k^{(m')}\} \subset C$ and $a_k^{(m)} \uparrow a_k, b_k^{(m')} \uparrow b_k$. From the continuity of finite measure and the definition of C, it follows

$$\begin{aligned}
\mu_1([a, b)) &= \lim_{m \uparrow \infty} \mu_1\left([a^{(m)}, b)\right) = \lim_{m \uparrow \infty} \lim_{m' \uparrow \infty} \mu_1\left(\left[a^{(m)}, b^{(m')}\right)\right) \\
&= \lim_{m \uparrow \infty} \lim_{m' \uparrow \infty} \mu_2\left(\left[a^{(m)}, b^{(m')}\right)\right) = \mu_2([a, b)).
\end{aligned}$$

□

6.2 Weak Convergence of Finite Measures

6.2.1 Definition and equivalent statements

Let (E, ρ) be a metric space and $\mathscr{E}$ be the Borel σ-algebra. Denote by $\mathfrak{M}$ the total of finite measures on $(E, \mathscr{E})$.

Lemma 6.8 (Regularity). *Let $\mu \in \mathfrak{M}$. Then $\forall A \in \mathscr{E}$,*

$$\mu(A) = \inf_{G \supset A,\, G \text{ is open}} \mu(G) = \sup_{C \subset A,\, C \text{ is closed}} \mu(C).$$

Proof. Let $\mathscr{C}$ be the class of all sets $A \in \mathscr{B}$ satisfying the desired equations. It suffices to prove (1) $\mathscr{C}$ contains all open sets, which is a π-system and (2) $\mathscr{C}$ is a λ-system.

To prove (1), let A be an open set. Then the first equation holds, and the second equation also holds if $A = E$. Now, let $A \neq E$ so that A^c is a nonempty closed set. By the triangle inequality, the distance function to A^c defined by $d(\cdot, A^c) := \inf\limits_{y \in A^c} \rho(\cdot, y)$ is Lipschitz continuous. Let $C_n = \left\{x \in E : d(x, A^c) \geqslant \frac{1}{n}\right\}$. Then C_n is closed and $C_n \subset A$. Since A is open, for any $x \in A$, there exists $n \geqslant 1$ such that $B(x, \frac{1}{n}) \subset A$. Thus, $d(x, A^c) \geqslant \frac{1}{n}$, so that $x \in C_n$. Therefore, $C_n \uparrow A (n \to \infty)$. By the continuity of μ, we obtain $\lim\limits_{n\to\infty} \mu(C_n) = \mu(A)$. This proves the second equation, so that $A \in \mathscr{C}$.

To prove (2), it suffices to show that $\mathscr{C}$ is a monotone class and closed under the proper difference. Let $\{A_n\}_{n \geqslant 1} \subset \mathscr{C}, A_n \uparrow A (n \to \infty)$. For every A_n, there exists an open set $G_n \supset A_n$ such that $|\mu(G_n) - \mu(A_n)| \leqslant 2^{-n}$; there also exists a closed set $C_n \subset A_n$ such that $|\mu(C_n) - \mu(A_n)| \leqslant 2^{-n}$. Then $\tilde{G}_n = \bigcup\limits_{m=n}^{\infty} G_m$ is an open set including A, while C_n is a closed set included in A. Moreover,

$$\begin{aligned}
\overline{\lim_{n\to\infty}} |\mu(\tilde{G}_n) - \mu(A)| &= \overline{\lim_{n\to\infty}} \left| \mu\left(\bigcup_{m=n}^{\infty} G_m\right) - \mu\left(\bigcup_{m=n}^{\infty} A_m\right) \right| \\
&\leqslant \overline{\lim_{n\to\infty}} \mu\left(\bigcup_{m=n}^{\infty} (G_m - A_m)\right) \\
&\leqslant \overline{\lim_{n\to\infty}} \sum_{m=n}^{\infty} 2^{-m} = 0,
\end{aligned}$$

and

$$\overline{\lim_{n\to\infty}} |\mu(C_n) - \mu(A)| = \overline{\lim_{n\to\infty}} |\mu(C_n) - \mu(A_n)|.$$

Therefore, $A \in \mathscr{C}$. Finally, let $A_1, A_2 \in \mathscr{C}$ with $A_1 \supset A_2$. It remains to be proved that $A_1 - A_2 \in \mathscr{C}$. For this, $\forall n \geqslant 1$, we take an open set $G_n \supset A_1$ and a closed set $C_n \subset A_2$ such that

$$|\mu(G_n) - \mu(A_1)| + |\mu(C_n) - \mu(A_2)| \leqslant \frac{1}{n}.$$

Then $G_n \setminus C_n$ is open, including $A_1 - A_2$, and

$$|\mu(A_1 - A_2) - \mu(G_n \setminus C_n)| \leqslant |\mu(G_n) - \mu(A_1)| + |\mu(C_n) - \mu(A_2)| \leqslant \frac{1}{n}.$$

So, $A := A_1 - A_2$ satisfies the first equation. Symmetrically, we can prove the second equation for $A := A_1 - A_2$, so that $A_1 - A_2 \in \mathscr{C}$. □

The above result shows that the class of open sets and that of closed sets are measure determined classes, i.e. two finite measures are equal if they coincide with each other on any one of these two classes. In the following, we prove that $C_b(E)$, the class of bounded continuous functions on E, is also a measure determined class. By the way, for later use, we also introduce the class $\mathscr{B}_b(E)$ of bounded measurable functions on E as well as $C_0(E)$ of continuous functions on E with compact supports.

Lemma 6.9. *Let $\mu_1, \mu_2 \in \mathfrak{M}$. If $\mu_1(f) = \mu_2(f)$ for $f \in C_b(E)$, then $\mu_1 = \mu_2$.*

Proof. By Lemma 6.8, it suffices to prove that $\mu_1(G) = \mu_2(G)$ for any open G. Let $g(x) = d(x, G^c)$ for $x \in E$. Then $g(x) > 0$ for $x \in G$ and g is Lipschitz. Set $h_n(r) = (nr) \wedge 1$. Then $h_n \circ g$ is Lipschitz and $h_n \circ g \uparrow 1_G$ $(n \uparrow \infty)$. From the monotone convergence theorem and $\mu_1(h_n \circ g) = \mu_2(h_n \circ g)$, it follows that $\mu_1(G) = \mu_2(G)$. □

Definition 6.10. Let $\{\mu_n\} \subset \mathfrak{M}$ and $\mu \in \mathfrak{M}$.

(1) We say that $(\mu_n)_{n\geqslant 1}$ converges uniformly to μ, denoted by $\mu_n \xrightarrow{u} \mu$, if

$$\sup_{A\in\mathscr{B}} |\mu_n(A) - \mu(A)| \to 0, \quad n \uparrow \infty,$$

equivalently,

$$\sup_{f\in\mathscr{B},|f|\leqslant 1} |\mu_n(f)-\mu(f)|\to 0, \quad n\uparrow\infty.$$

(2) We say that $(\mu_n)_{n\geqslant 1}$ converges strongly to μ, denoted by $\mu_n \xrightarrow{s} \mu$, if

$$\lim_{n\to\infty}\mu_n(A)=\mu(A), \quad \forall A\in\mathscr{B},$$

equivalently, $\mu_n(f)\to\mu(f)$ for every $f\in\mathscr{B}_b$.

(3) We call $(\mu_n)_{n\geqslant 1}$ convergent weakly to μ, denoted by $\mu_n \xrightarrow{w} \mu$, if $\mu_n(f)\to\mu(f)$ for every $f\in C_b(E)$.

(4) $(\mu_n)_{n\geqslant 1}$ is called convergent vaguely to μ, denoted by $\mu_n \xrightarrow{v} \mu$, if $\mu_n(f)\to\mu(f)$ for every $f\in C_0(E)$.

Definition 6.11. A set $A\in\mathscr{E}$ is called μ-continuous if $\mu(\partial A)=0$.

Following are some equivalent characterizations on the weak convergence.

Theorem 6.12. *Let $\mu_n,\mu\in\mathfrak{M}$ $(n\geqslant 1)$. The following statements are equivalent:*

(1) *$\mu_n(f)\to\mu(f)$ for every $f\in C_b(E)$.*
(2) *$\mu_n(f)\to\mu(f)$ for every bounded uniformly continuous function f.*
(3) *$\mu_n(f)\to\mu(f)$ for every bounded Lipschitz continuous function f.*
(4) *$\varliminf_{n\to\infty}\mu_n(G)\geqslant\mu(G)$ for every open $G\subset E$, and $\mu_n(E)\to\mu(E)$.*
(5) *$\varlimsup_{n\to\infty}\mu_n(C)\leqslant\mu(C)$ for every closed $C\subset E$, and $\mu_n(E)\to\mu(E)$.*
(6) *$\mu_n(A)\to\mu(A)$ for every μ-continuous set A.*

Proof. (1) $\Rightarrow$ (2) $\Rightarrow$ (3) and (4) $\Leftrightarrow$ (5) are obvious.
(3) $\Rightarrow$ (5). Let $C\subset E$ be a closed set, and define

$$f_m(x)=\frac{1}{1+md(x,C)}, \quad x\in E, \quad m\geqslant 1.$$

Then f_m is Lipschitz and $1 \geqslant f_m \downarrow \mathbf{1}_C$ as $m \uparrow \infty$. By (3) and the dominated convergence theorem, we obtain

$$\begin{aligned}\mu(C) &= \lim_{m\to\infty}\int_E f_m\,\mathrm{d}\mu = \lim_{m\to\infty}\lim_{n\to\infty}\int_E f_m\,\mathrm{d}\mu_n\\ &\geqslant \overline{\lim_{n\to\infty}}\,\mu_n(C).\end{aligned}$$

(4) and (5) $\Rightarrow$ (6). Let A be a μ-continuous set. Then $\mu(A) = \mu(\bar{A}) = \mu(A^\circ)$, where $\bar{A}$ and A° are the closure and interior of A, respectively. This together with (4) and (5) yields

$$\begin{aligned}\mu(A) &= \mu(A^\circ) \leqslant \underline{\lim_{n\to\infty}}\,\mu_n(A^\circ) \leqslant \underline{\lim_{n\to\infty}}\,\mu_n(A),\\ \mu(A) &= \mu(\bar{A}) \geqslant \overline{\lim_{n\to\infty}}\,\mu_n(\bar{A}) \geqslant \overline{\lim_{n\to\infty}}\,\mu_n(A).\end{aligned}$$

So, (6) holds.

(6) $\Rightarrow$ (1). $\forall f \in C_b(E)$, we intend to find a sequence of simple functions $\{f_n\}_{n\geqslant 1}$ generated by μ-continuous sets such that $f_n \to f$ uniformly as $n \to \infty$. Since μ is finite, the set

$$D := \{a \in \mathbb{R} : \mu(\{f = a\}) > 0\}$$

is at most countable. Thus, we may find a constant $c > ||f||_\infty + 1$ such that $\pm c \in D^c$, and a sequence of partitions

$$I_n := \{-c = r_0 < r_1 < \cdots < r_n < r_{n+1} = c\}, \quad n \geqslant 2$$

such that

$$\{r_i\} \subset D^c,\ \delta(I_n) := \max_{1\leqslant k\leqslant n+1}(r_k - r_{k-1}) \to 0.$$

Let

$$f_n = \sum_{i=1}^{n-1} r_i \mathbf{1}_{\{r_i \leqslant f < r_{i+1}\}}.$$

Then

$$||f_n - f||_\infty \leqslant \delta(I_n) \to 0, \quad n \to \infty,$$

so that

$$|\mu(f) - \mu_m(f)| \leqslant |\mu(f) - \mu(f_n)| + |\mu(f_n) - \mu_m(f_n)| + |\mu_m(f_n) - \mu_m(f)|$$

$$\leqslant \delta(I_n)\,(\mu(E) + \mu_m(E)) + \sum_{i=1}^{n-1} r_i\,|\mu(r_i \leqslant f < r_{i+1}) - \mu_m(r_i \leqslant f < r_{i+1})|.$$

Noting that $\{r_i\} \subset D^c$ implies that each set $\{r_i \leqslant f < r_{i+1}\}$ is μ-continuous, by (6), we may let first $m \uparrow \infty$ then $n \uparrow \infty$ to derive (1). □

6.2.2 Tightness and weak compactness

The topology induced by weak convergence on $\mathfrak{M}$ is called the weak topology. In this part, we characterize the weak compactness (i.e. compactness in the weak topology) for subsets of $\mathfrak{M}$. We first consider a simple case where E is a compact metric space. In this case, the relatively weak compactness is equivalent to the boundedness, which is well known for subsets in the Euclidean space. Recall that $\mathfrak{M}' \subset \mathfrak{M}$ is called bounded if $\sup_{\mu\in\mathfrak{M}'} \mu(E) < \infty$.

Theorem 6.13. *Let (E, ρ) a compact metric space. If $\{\mu_n\} \subset \mathfrak{M}$ is bounded, then there exists a subsequence $\{\mu_{n_k}\}$ such that $\mu_{n_k} \xrightarrow{w} \mu$ as $k \to \infty$ for some $\mu \in \mathfrak{M}$.*

Proof. Since E is compact, $C(E)$ is a Polish space under uniform norm. Let $\{f_n\}_{n\geqslant1}$ be a dense subset of $C(E)$. Since $\{\mu_n\}_{n\geqslant1}$ is bounded, for each $m \geqslant 1$, $\{\mu_n(f_m)\}_{n\geqslant1}$ is bounded in $\mathbb{R}$ and hence has a convergent subsequence. By the diagonal principle, we may find a subsequence $n_k \uparrow \infty$ as $k \uparrow \infty$ and a sequence $\{\alpha_m\}_{m\geqslant1} \subset \mathbb{R}$ such that

$$\lim_{k\to\infty} \mu_{n_k}(f_m) = \alpha_m, \quad m \geqslant 1.$$

Since $\{f_n\}_{n\geqslant1}$ is dense in $C(E)$, for any $f \in C(E)$ and $\varepsilon > 0$, there exists $m_0 \geqslant 1$ such that $||f_{m_0} - f||_\infty \leqslant \varepsilon$. So,

$$\begin{aligned}
&|\mu_{n_k}(f) - \mu_{n_l}(f)| \\
&\quad \leqslant |\mu_{n_k}(f - f_{m_0})| + |\mu_{n_l}(f - f_{m_0})| + |\mu_{n_k}(f_{m_0}) - \mu_{n_l}(f_{m_0})| \\
&\quad \leqslant 2\varepsilon C + |\mu_{n_k}(f_{m_0}) - \mu_{n_l}(f_{m_0})|.
\end{aligned}$$

By letting first $k, l \to \infty$ then $\varepsilon \to 0$, we obtain

$$\lim_{l,k\to\infty} |\mu_{n_k}(f) - \mu_{n_l}(f)| = 0.$$

Then $\{\mu_{n_k}(f)\}$ is a Cauchy sequence and there exists $\alpha(f) \in \mathbb{R}$ such that $\mu_{n_k}(f) \to \alpha(f)$. It is clear that $\alpha : C(E) \to \mathbb{R}$ is a nonnegative bounded linear functional. By Riesz–Markov–Kakutani theorem [19, Theorem IV.14], there exists unique $\mu \in \mathfrak{M}$ such that $\mu(f) = \alpha(f)$. Therefore, $\mu_{n_k} \xrightarrow{w} \mu$. □

When E is not compact, the above results remains true if the bounded sequence $\{\mu_n\}_{n\geqslant 1}$ is supported on a compact set $K \subset E$, i.e. $\mu_n(K^c) = 0, n \geqslant 1$. In general, we may extend the result to bounded $\{\mu_n\}_{n\geqslant 1}$ asymptotically supported on compact sets. This leads to the notion of tightness.

Definition 6.14. A bounded subset $\mathfrak{M}'$ of $\mathfrak{M}$ is called tight if for any $\varepsilon > 0$, there exists a compact set $K \subset E$ such that

$$\sup_{\mu\in\mathfrak{M}'} \mu(K^c) < \varepsilon.$$

Theorem 6.15 (Prohorov's theorem). *Let (E, ρ) be a metric space and let $\{\mu_n\}_{n\geqslant 1} \subset \mathfrak{M}$ be bounded.*

(1) *If there exists a sequence of compact sets $\{K_m\}_{m\geqslant 1}$ such that $K_m \uparrow E$, then $\{\mu_n\}_{n\geqslant 1}$ has a vague convergent subsequence.*
(2) *If $\{\mu_n\}_{n\geqslant 1}$ is tight, then it has a weak convergent subsequence.*

Proof. Let $\{K_m\}_{m\geqslant 1}$ be a sequence of increasing compact subsets of E. Given m, there exist a subsequence $\{\mu_{m_k}\}$ and a finite measure $\mu^{(m)}$ on K_m such that

$$\mu_{m_n}|_{K_m} \xrightarrow{w} \mu^{(m)} \ (n \to \infty),$$

where $\mu_{m_n}|_{K_m}$ is the restriction of μ_{m_n} on K_m. By the diagonal principle, there exists a subsequence $\{\mu_{n_k}\}$ such that

$$\mu_{n_k}|_{K_m} \xrightarrow{w} \mu^{(m)} \ (k \to \infty), \quad m \geqslant 1.$$

Clearly,

$$\mu^{(m+1)}(A \cap K_{m+1}) \geqslant \mu^{(m)}(A \cap K_m), \quad \forall A \in \mathscr{B}.$$

Indeed, for any closed set A, let

$$h_l = \frac{1}{1 + ld(x, A)} \quad l \geqslant 1.$$

Then

$$\begin{aligned}
\mu^{(m+1)}(A \cap K_{m+1}) &= \lim_{l \to \infty} \mu^{(m+1)}(h_l \mathbf{1}_{K_{m+1}}) \\
&= \lim_{l \to \infty} \lim_{k \to \infty} \mu_{n_k}(h_l \mathbf{1}_{K_{m+1}}) \\
&\geqslant \lim_{l \to \infty} \lim_{n_k \to \infty} \mu_{n_k}(h_l \mathbf{1}_{K_m}) \\
&= \mu^{(m)}(A \cap K_m).
\end{aligned}$$

Thus, limit $\mu(A) = \lim_{m \to \infty} \mu^{(m)}(A \cap K_m)$ exists for any $A \in \mathscr{B}$, so that $\mu \in \mathfrak{M}$ and $\mu^{(m)}(f \mathbf{1}_{K_m}) \to \mu(f)$ for any $f \in C_b(E)$.

(1) Since $K_m \uparrow E$, for any $f \in C_0(E)$, there exists $m_0 \geqslant 1$ such that $\text{supp} f \subset K_m$ for all $m \geqslant m_0$. Thus,

$$\lim_{k \to \infty} \mu_{n_k}(f) = \lim_{m \to \infty} \mu^{(m)}(f) = \mu(f).$$

(2) Up to a subsequence, we may assume that $\sup_{n \geqslant 1} \mu_n(K_m^c) \leqslant 1/m$ for $m \geqslant 1$. Then $\forall f \in C_b(E)$,

$$\begin{aligned}
&|\mu_{n_k}(f) - \mu(f)| \\
&\leqslant |\mu_{n_k}(f \mathbf{1}_{K_m}) - \mu^{(m)}(f \mathbf{1}_{K_m})| + |\mu_{n_k}(f - f \mathbf{1}_{K_m})| + |\mu(f) - \mu^{(m)}(f \mathbf{1}_{K_m})| \\
&\leqslant \frac{1}{m} ||f||_\infty + |\mu_{n_k}(f \mathbf{1}_{K_m}) - \mu^{(m)}(f \mathbf{1}_{K_m})| + |\mu(f) - \mu^{(m)}(f \mathbf{1}_{K_m})|.
\end{aligned}$$

By first letting $k \uparrow \infty$ and then $m \uparrow \infty$, we prove $\mu_{n_k} \xrightarrow{w} \mu$. □

Theorem 6.16. *Let E be a Polish space. Then a subset of $\mathfrak{M}$ is weak relatively compact if and only if it is tight.*

Proof. By Prohorov's theorem, we only need to prove the necessity. Let $\mathfrak{M}' \subset \mathfrak{M}$ be relatively compact, we intend to prove the tightness. To this end, we first observe that

$$\lim_{n\to\infty} \sup_{\mu\in\mathfrak{M}'} \mu(G_n^c) = 0 \tag{6.1}$$

holds for any increasing open sets $G_n \uparrow E$. To see this, for any $n \geqslant 1$, we take $\mu_n \in \mathfrak{M}'$ such that

$$\mu_n(G_n^c) \geqslant \sup_{\mu\in\mathfrak{M}'} \mu(G_n^c) - 1/n.$$

Since $\mathfrak{M}'$ is weak relatively compact, there exist $\mu_0 \in \mathfrak{M}$ and a subsequence $\mu_{n_k} \xrightarrow{w} \mu_0$. Combining this with the increasing property of G_n, we obtain

$$\begin{aligned}\lim_{n\to\infty} \sup_{\mu\in\mathfrak{M}'} \mu(G_n^c) &= \lim_{k\to\infty} \sup_{\mu\in\mathfrak{M}'} \mu(G_{n_k}^c) \leqslant \overline{\lim}_{k\to\infty} \mu_{n_k}(G_{n_k}^c)\\ &\leqslant \lim_{m\to\infty} \overline{\lim}_{k\to\infty} \mu_{n_k}(G_m^c) \leqslant \lim_{m\to\infty} \mu_0(G_m^c) = 0.\end{aligned}$$

So, (6.1) holds.

Since E is separable, $\forall m \geqslant 1, \exists\{x_{m,j}\}$ such that $E = \bigcup_{j=1}^{\infty} B(x_{m,j}, 2^{-m})$. Let $G(n,m) = \bigcup_{j=1}^{n} B(x_{m,j}, 2^{-m})$. Then $K_\varepsilon := \bigcap_{m=1}^{\infty} G(N(\varepsilon,m),m)$ is completely bounded, so that by Hausdorff's theorem, $\bar{K}_\varepsilon$ is compact. Since $G(n,m) \uparrow E$ $(n \uparrow \infty)$, by (6.1) with $G_n = G(n,m)$, we conclude that $\forall \varepsilon > 0, \exists N(\varepsilon,m) \geqslant 1$ such that

$$\sup_{\mu\in\mathfrak{M}'} \mu(G(n,m)^c) \leqslant \frac{\varepsilon}{2^m},\ n \geqslant N(\varepsilon,m),$$

and

$$\mu(\bar{K}_\varepsilon^c) \leqslant \sum_{r=1}^{\infty} \mu(G(N,r)^c) \leqslant \sum_{r=1}^{\infty} \frac{\varepsilon}{2^r} = \varepsilon,\ \mu \in \mathfrak{M}'. \qquad \square$$

When $E = \mathbb{R}^d$, we have the following one more equivalent statement for the weak convergence using continuous intervals in place of continuous sets.

Proposition 6.17. *Let* $E = \mathbb{R}^d$. *Then* $\mu_n \xrightarrow{w} \mu$ *if and only if* $\mu_n(\mathbb{R}^d) \to \mu(\mathbb{R}^d)$ *and* $\mu_n([a,b)) \to \mu([a,b))$ *for any finite* μ*-continuous interval* $[a,b)$.

Proof. We only need to prove sufficiency. If μ_n is not weakly convergent to μ, then there exist $\delta > 0, f \in C_b(\mathbb{R}^d)$ and subsequence $n_k \to \infty$ such that

$$|\mu_{n_k}(f) - \mu(f)| \geqslant \delta, \quad k \geqslant 1. \tag{6.2}$$

By Lemmas 6.5 and 6.6, there exists a sequence of μ-continuous intervals $I_m \uparrow \mathbb{R}^d$. Then $\forall \varepsilon > 0$, there exists $m \geqslant 1$ such that $\mu(I_m^c) \leqslant \varepsilon/2$. Since $\mu_n(I_m) \to \mu(I_m)$ and $\mu_n(\mathbb{R}^d) \to \mu(\mathbb{R}^d)$ when $n \to \infty$, we have $\overline{\lim_{n\to\infty}} \mu_n(I_m^c) \leqslant \varepsilon/2$. Thus, there exists $n_0 \geqslant 1$ such that $\forall n \geqslant n_0$, $\mu_n(I_m^c) < \varepsilon$. Moreover, take compact set K_1 such that $\mu_n(K_1^c) < \varepsilon$ for $\forall n \leqslant n_0$. Then $K = K_1 \cup \bar{I}_m$ is compact and satisfies $\mu_n(K^c) < \varepsilon, \forall n \geqslant 1$. Thus, $\{\mu_{n_k}\}$ is tight, so there exist a subsequence n_k' and a finite measure μ' such that $\mu_{n_k'} \xrightarrow{w} \mu'$. Combining this with the condition that $\mu_n([a,b)) \to \mu([a,b))$ for μ-continuous intervals $[a,b)$, we see that μ' and μ are equal on their common continuous intervals. By Proposition 6.7, we have $\mu' = \mu$, which contradicts to (6.2) since $\mu_{n_k'} \xrightarrow{w} \mu'$. □

6.3 Characteristic Function and Weak Convergence

In this section, we first identify the weak convergence for finite measures on $\mathbb{R}^n$ by using the convergence of characteristic functions and then prove that a complex function on $\mathbb{R}^n$ is a characteristic function if and only if it is continuous and nonnegative definite.

Theorem 6.18. *Let $\{\mu_k, \mu\}_{k\geqslant 1}$ be finite measures on $\mathbb{R}^n$. Then $\mu_k \xrightarrow{w} \mu$ $(k \to \infty)$ if and only if $f_{\mu_k} \to f_\mu$ pointwise.*

By the dominated convergence theorem, the necessity is obvious. The sufficiency follows from Theorem 6.22 on the convergence of integral characteristic functions.

Definition 6.19. Let f_μ be the characteristic function of a finite measure μ. The indefinite integral of f_μ

$$\tilde{f}_\mu(u_1, \ldots, u_n) = \int_0^{u_1} \cdots \int_0^{u_n} f_\mu(t_1, \ldots, t_n)\, \mathrm{d}t_1 \cdots \mathrm{d}t_n, \quad u \in \mathbb{R}^n$$

is called the integral characteristic function of μ, where $\int_0^{u_i} = -\int_{u_i}^0$ if $u_i < 0$.

Since f_μ is continuous, f_μ and $\tilde{f}_\mu$ determine each other.

Lemma 6.20. *The integral characteristic function of μ satisfies*

$$\tilde{f}_\mu(u_1, \ldots, u_n) = \int_{\mathbb{R}^n} \prod_{k=1}^n \frac{e^{iu_k x_k} - 1}{ix_k} \mu(\mathrm{d}x_1, \ldots, \mathrm{d}x_n),\ u_1, \ldots, u_n \in \mathbb{R}.$$

Proof. By the definition and Fubini's theorem, for $u = (u_1, \ldots, u_n) \in \mathbb{R}^n$,

$$\begin{aligned}
\tilde{f}_\mu(u) &= \int_0^{u_1} \cdots \int_0^{u_n} \int_{\mathbb{R}^n} \mathrm{e}^{\mathrm{i}\langle t,x\rangle} \mu(\mathrm{d}x)\, \mathrm{d}t \\
&= \int_{\mathbb{R}^n} \mu(\mathrm{d}x) \int_{[0,u]} \mathrm{e}^{\mathrm{i}\langle t,x\rangle}\, \mathrm{d}t \\
&= \int_{\mathbb{R}^n} \prod_{k=1}^n \frac{\mathrm{e}^{\mathrm{i}u_k x_k} - 1}{\mathrm{i}x_k} \mu(\mathrm{d}x_1, \ldots, \mathrm{d}x_n).
\end{aligned}$$

□

Let

$$F(x, u) = \prod_{k=1}^n \frac{\mathrm{e}^{\mathrm{i}\, u_k x_k} - 1}{\mathrm{i}\, x_k},\ x, u \in \mathbb{R}^n.$$

Then for given u, $\lim_{|x|\to\infty} F(x, u) = 0$. Thus, $F(\cdot, u)$ can be uniformly approximated by continuous functions with compact supports.

Theorem 6.21. *Let $\{\mu_k\}_{k\geqslant 1}$ be bounded measures on $\mathbb{R}^n$. If $\tilde{f}_{\mu_k} \to \tilde{g}$ for some function $\tilde{g}$, then there exists a finite measure μ such that $\mu_k \xrightarrow{v} \mu$ and $\tilde{g} = \tilde{f}_\mu$.*

Proof. By Theorem 6.15, there exists a subsequence $\{\mu_{n_k}\}_{k\geqslant 1}$ of $\{\mu_k\}_{k\geqslant 1}$ which converges vaguely to some finite measure μ. Since a finite measure is determined by its integral characteristic function, we only need to prove $\tilde{f}_\mu = \tilde{g}$. Since $\tilde{f}_\mu(u) = \mu(F(u, \cdot))$, and $F(u, \cdot)$ can be uniformly approximated by continuous functions with compact supports, it is clear that $\mu_{n_k} \xrightarrow{v} \mu$ and $\tilde{f}_{\mu_k} \to \tilde{g}$ imply $\tilde{f}_\mu = \tilde{g}$. □

Theorem 6.22. *Let $\{\mu_k\}_{k\geqslant 1}$ be bounded measures on $\mathbb{R}^n$ such that $f_{\mu_k} \to g$ for some function g continuous at 0. Then there exists a finite measure μ such that $\mu_k \xrightarrow{w} \mu$ and $f_\mu = g$.*

Proof. By the dominated convergence theorem, $f_{\mu_k} \to g$ implies $\tilde{f}_{\mu_k} \to \tilde{g}$. By Theorem 6.21, Proposition 6.17 and Exercise 9, it suffices to prove $\mu_k(\mathbb{R}^n) \to \mu(\mathbb{R}^n)$. Since $\tilde{g} = \tilde{f}_\mu$, $g = f_\mu$ $\mathrm{d}x$-a.e., and since both g and f_μ are continuous at 0,

$$\mu(\mathbb{R}^n) = f_\mu(0) = g(0) = \lim_{k\to\infty} f_k(0) = \lim_{k\to\infty} \mu_k(\mathbb{R}^n). \qquad \square$$

In the following, we introduce two important applications of Theorem 6.18.

Theorem 6.23 (Law of large numbers). *Let $\{\xi_n\}$ be i.i.d. random variables with $\mathbb{E}\xi_n = a \in \mathbb{R}$. Then*

$$\frac{1}{n}\sum_{k=1}^n \xi_k \xrightarrow{\mathbb{P}} a.$$

Proof. (1) It suffices to prove the characteristic functions f_n of $\eta_n := \frac{1}{n}\sum\limits_{k=1}^n (\xi_k - a)$ satisfy $f_n(t) \to 1$. In fact, if $f_n(t) \to 1$, then by Theorem 6.18, we have $\mathbb{P}_{\eta_n} \xrightarrow{w} \delta_0$ (probability with total mass at 0). Since $(-\varepsilon, \varepsilon)$ is δ_0-continuous for any $\varepsilon > 0$, there holds

$$\underline{\lim}_{n\to\infty} \mathbb{P}\left(\left|\frac{1}{n}\sum_{k=1}^n \xi_k - a\right| < \varepsilon\right) = \underline{\lim}_{n\to\infty} \mathbb{P}_{\eta_n}((-\varepsilon,\varepsilon)) = \delta_0((-\varepsilon,\varepsilon)) = 1,$$

which implies

$$\lim_{n\to\infty} \mathbb{P}\left(\left|\frac{1}{n}\sum_{k=1}^n \xi_k - a\right| \geqslant \varepsilon\right) = 0.$$

(2) Let $\xi_n' = \xi_n - a$. Then $\eta_n = \frac{1}{n}\sum\limits_{k=1}^n \xi_k'$, so

$$f_n(t) = \prod_{k=1}^n f_{\xi_k'}(t/n) = [f(t/n)]^n,$$

where $f = f_{\xi'_k}$. Since $\mathbb{E}\xi'_k = 0$, Taylor's expansion gives

$$f_n(t) = \left[\mathbb{E}e^{it\xi'_k/n}\right]^n = (1 + o(1/n))^n, \quad t \in \mathbb{R}.$$

Thus,

$$\lim_{n\to\infty} \log f_n(t) = \lim_{n\to\infty} \log\left[(1 + o(1/n))^n\right] = \lim_{n\to\infty} n \log(1 + o(1/n)) = 0.$$

Therefore, $\lim_{n\to\infty} f_n(t) = 1$. □

Theorem 6.24 (Central limit theorem). *Let $\{\xi^{(k)}\}_{k\geqslant 1}$ be a sequence of n-dimensional i.i.d. random variables with invertible correlation matrix D and $\mathbb{E}\xi^{(k)} = m \in \mathbb{R}^n$. Then $\forall x \in \mathbb{R}^n$,*

$$\lim_{N\to\infty} \mathbb{P}\left(\frac{1}{\sqrt{N}}\sum_{k=1}^{N}(\xi^{(k)} - m) < x\right) = \frac{1}{(2\pi)^{n/2}|D|^{1/2}} \int_{(-\infty,x)} e^{-\frac{1}{2}\langle t, D^{-1}t\rangle}\, dt.$$

Proof. Let $\eta^{(k)} = \xi^{(k)} - m$. Then $\{\eta^{(k)}\}$ are i.i.d with zero mean. Let f be the characteristic function of $\eta^{(k)}$. Then the characteristic function of $\frac{1}{\sqrt{N}}\sum_{k=1}^{N}\eta^{(k)}$ is

$$f_N(t) = \left[f(t/\sqrt{N})\right]^N, \quad t \in \mathbb{R}^n.$$

Since $\mathbb{E}\eta^{(k)} = 0$, Taylor's expansion shows

$$f(t/\sqrt{N}) = 1 - \frac{1}{2N}\langle t, Dt\rangle + o(1/N), \quad t \in \mathbb{R}^n,$$

so that

$$\log f(t/\sqrt{N}) = -\frac{1}{2N}\langle t, Dt\rangle + o(1/N).$$

Thus,

$$\lim_{N\to\infty} \log f_N(t) = -\frac{1}{2}\langle t, Dt\rangle, \quad t \in \mathbb{R}^n,$$

so that

$$\lim_{N\to\infty} f_N(t) = e^{-\frac{1}{2}\langle t, Dt\rangle}.$$

By Theorem 6.18, this implies that $\left\{\frac{1}{\sqrt{N}}\sum_{k=1}^{N}\eta^{(k)}\right\}$ converges in distribution to $N(0, D)$, the centered normal distribution with covariance D. □

6.4 Characteristic Function and Nonnegative Definiteness

Let μ be a finite measure on $\mathbb{R}^n$ with characteristic function f_μ. Clearly, $\forall m \geqslant 1$, $\alpha_1, \ldots, \alpha_m \in \mathbb{C}$, and $t^{(1)}, \ldots, t^{(m)} \in \mathbb{R}^n$, we have

$$\sum_{j,k=1}^{m} f_\mu\left(t^{(j)} - t^{(k)}\right)\alpha_j\bar{\alpha}_k = \int_{\mathbb{R}^n}\left|\sum_{k=1}^{m}\alpha_k \mathrm{e}^{\mathrm{i}\langle t^{(k)}, x\rangle}\right|^2 \mu(\mathrm{d}x) \geqslant 0.$$

A function having this property is called a nonnegative definite function, and this property is called the nonnegative definiteness. In this section, we will prove that a function on $\mathbb{R}^n$ is the characteristic function of a finite measure if and only if it is continuous and nonnegative definite. To this end, we first observe that a nonnegative function has some properties of characteristic functions.

Property 6.25. If f is a nonnegative definite function, then $f(0) \geqslant 0, f(-t) = \bar{f}(t)$ and $|f(t)| \leqslant f(0)$.

Proof. Let $m = 2, t^{(1)} = 0, t^{(2)} = t, \alpha_1 = 1, \alpha_2 \in \mathbb{C}$. By the nonnegative definiteness, we have

$$f(0)\left[1 + |\alpha_2|^2\right] + f(t)\alpha_2 + f(-t)\bar{\alpha}_2 \geqslant 0.$$

(1) Let $\alpha_2 = 0$. Then $f(0) \geqslant 0$.

(2) Let $\alpha_2 = 1$. Then $2f(0) + f(-t) + f(t) \geqslant 0$, so that $\mathrm{Im} f(t) = -\mathrm{Im} f(-t)$. Moreover, taking $\alpha_2 = \mathrm{i}$, we obtain $2f(0) + \mathrm{i}(f(t) - f(-t)) \geqslant 0$, so that $\mathrm{Re} f(t) = \mathrm{Re} f(-t)$. In conclusion, $f(-t) = \bar{f}(t)$.

(3) For $f(t) \neq 0$ and $\alpha_2 := -\bar{f}(t)/|f(t)|$, we obtain $2f(0) \geqslant 2|f(t)|$, so that $f(0) \geqslant |f(t)|$. □

Lemma 6.26. *Let $T_c = \{kc : k \in \mathbb{Z}^n\}, c > 0$. If f is a nonnegative definite function, then there exists a finite measure μ on $\mathbb{R}^n$ such that*

$$\mu(\mathbb{R}^n) = \mu([-\pi/c, \pi/c]^n) = f(0), \quad f_\mu(t) = f(t), \quad \forall t \in T_c.$$

Proof. $\forall m \geqslant 1$, by the nonnegative definiteness of f, we obtain

$$\begin{aligned} 0 &\leqslant \frac{1}{m^n} \sum_{j_1,\dots,j_n,k_1,\dots,k_n=0}^{m-1} f(c(j-k))\mathrm{e}^{-\mathrm{i}\,c\langle j-k,x\rangle} \\ &= \sum_{r_1,\dots,r_n=-m}^{m} \left[\prod_{\ell=1}^{n}\left(1-\frac{|r_\ell|}{m}\right)\right] f(cr)\mathrm{e}^{-\mathrm{i}\,c\langle r,x\rangle} =: G_m(x). \end{aligned}$$

Let

$$\mu_m(\mathrm{d}x) = \left(\frac{c}{2\pi}\right)^n G_m(x)\mathbf{1}_{[-\frac{\pi}{c},\frac{\pi}{c}]^n}(x)\,\mathrm{d}x.$$

Then

$$\begin{aligned} \mu_m(\mathbb{R}^n) &= \mu_m\left(\left[-\frac{\pi}{c},\frac{\pi}{c}\right]^n\right) \\ &= \sum_{r_1,\dots,r_n=-m}^{m} \left[\prod_{\ell=1}^{n}\left(1-\frac{|r_\ell|}{m}\right)\right] f(cr) \left[\prod_{\ell=1}^{n}\frac{c}{2\pi}\int_{-\frac{\pi}{c}}^{\frac{\pi}{c}} \mathrm{e}^{\mathrm{i}\,cr_\ell x_\ell}\,\mathrm{d}x_\ell\right] \\ &= f(0). \end{aligned}$$

Let f_m be the characteristic function of μ_m. Then

$$\begin{aligned} f_m(ck) &= \left(\frac{c}{2\pi}\right)^n \int_{[-\frac{\pi}{c},\frac{\pi}{c}]^n} \mathrm{e}^{\mathrm{i}\,c\langle k,x\rangle} G_m(x)\,\mathrm{d}x \\ &= \sum_{r_1,\dots,r_n=-m}^{m} \left[\prod_{\ell=1}^{n}\left(1-\frac{|r_\ell|}{m}\right)\right] f(cr) \left[\prod_{\ell=1}^{n}\frac{c}{2\pi}\int_{-\frac{\pi}{c}}^{\frac{\pi}{c}} \mathrm{e}^{\mathrm{i}\,c(k_\ell-r_\ell)x_\ell}\,\mathrm{d}x_\ell\right] \\ &= f(ck)\prod_{\ell=1}^{n}\left(1-\frac{|r_\ell|}{m}\right) \to f(ck) \quad (m\to\infty). \end{aligned}$$

Since $\{\mu_m\}_{m\geqslant 1}$ is tight, there exist μ and a subsequence μ_{m_k} such that $\mu_{m_k} \xrightarrow{w} \mu(k \to \infty)$. Then

$$\mu(\mathbb{R}^n) = \mu\left(\left[-\frac{\pi}{c}, \frac{\pi}{c}\right]^n\right) = f(0)$$

and

$$f_\mu(ck) = \lim_{m\to\infty} f_m(ck) = f(ck). \qquad \square$$

Theorem 6.27. *If f is a continuous and nonnegative definite function on $\mathbb{R}^n$, then it is the characteristic function of a finite measure.*

Proof. By Lemma 6.26, there exists a sequence of finite measures $\{\mu_m\}_{m\geqslant 1}$ such that

$$\mu_m(\mathbb{R}^n) = \mu_m\left([-m\pi, m\pi]^n\right) = f(0),$$

and their characteristic functions f_m satisfy $f_m(t) = f(t), t \in \frac{1}{m}\mathbb{Z}^n$. $\forall t \in \mathbb{R}^n$, take $\left\{t^{(m)}\right\}_{m\geqslant 1} \subset T_{1/m}$ such that $|t_k - t_k^{(m)}| \leqslant 1/m, 1 \leqslant k \leqslant n, m \geqslant 1$. Thus, by the continuity of f and $f\left(t^{(m)}\right) = f_m\left(t^{(m)}\right)$, we have

$$f(t) = \lim_{m\to\infty} f\left(t^{(m)}\right) = \lim_{m\to\infty} f_m\left(t^{(m)}\right).$$

From this and Theorem 6.18, it suffices to prove

$$\lim_{m\to\infty}\left|f_m(t) - f_m\left(t^{(m)}\right)\right| = 0. \tag{6.3}$$

For this, we use the increment inequality (Property 6.2) to derive

$$\begin{aligned}
&\left|f_m(t) - f_m\left(t^{(m)}\right)\right| \\
&\leqslant \sum_{i=0}^{n-1}\left|f_m\left(t_1,\ldots,t_i,t_{i+1}^{(m)},\ldots,t_n^{(m)}\right) - f_m\left(t_1,\ldots,t_{i+1},t_{i+2}^{(m)},\ldots,t_n^{(m)}\right)\right| \\
&\leqslant \sum_{i=0}^{n-1}\sqrt{2f(0)(f(0) - \mathrm{Re} f_m\left(e_i\left(t_i - t_i^{(m)}\right)\right)}, \qquad (6.4)
\end{aligned}$$

where $e_i \in \mathbb{R}^n$ is a unit vector with ith being 1. Since for $x_i \in [-m\pi, m\pi]$, $|(t_i - t_i^{(m)})x_i| \leqslant \pi$, and $\cos\theta$ is decreasing in $|\theta|$ on $\theta \in [-\pi, \pi]$, we have

$$\begin{aligned}
0 &\leqslant f(0) - \mathrm{Re} f_m\left(e_i\left(t_i - t_i^{(m)}\right)\right) \\
&= \int_{[-m\pi, m\pi]^n} \left(1 - \cos\left[\left(t_i - t_i^{(m)}\right) x_i\right]\right) \mu_m(\mathrm{d}x) \\
&\leqslant \int_{[-m\pi, m\pi]^n} \left(1 - \cos\frac{x_i}{m}\right) \mu_m(\mathrm{d}x) \\
&= f(0) - \mathrm{Re} f_m\left(\frac{e_i}{m}\right).
\end{aligned}$$

Combining this with (6.4) and the continuity of f, we prove (6.3). □

6.5 Exercises

1. Prove that the characteristic function f_μ of a finite measure μ on $\mathbb{R}^n$ has the following properties:

   (1) $f_\mu(0) = \mu(\mathbb{R}^n)$, (2) $|f_\mu(t)| \leqslant f(0)$, (3) $\bar{f}_\mu(t) = f(-t)$.

2. Prove Property 6.2(2).

3. A finite measure μ on $(\mathbb{R}, \mathscr{B})$ is called symmetric if $\mu(-\infty, x) = \mu(x, \infty)$ for any $x \geqslant 0$. Prove the following:

   (a) μ is symmetric if and only if $\mu(A) = \mu(-A)$, $A \in \mathscr{B}$, where $-A = \{x : -x \in A\}$;

   (b) μ is symmetric if and only if its characteristic function is a real function.

4. Let μ be a finite measure on $\mathbb{R}^n$ such that $\int_{\mathbb{R}} |f_\mu(t)|\,\mathrm{d}t < \infty$. Prove that μ is absolutely continuous with respect to $\mathrm{d}x$ and

$$\frac{\mu(\mathrm{d}x)}{\mathrm{d}x} = \frac{1}{2\pi} \int_{\mathbb{R}} \mathrm{e}^{-\mathrm{i}\,tx} \phi(t)\,\mathrm{d}t.$$

5. Prove Proposition 6.3.

6. Let $\{\xi_n, \xi\}_{n\geqslant 1}$ be centered normal random variables with variances $\{\sigma_n^2, \sigma^2\}_{n\geqslant 1}$. If $\xi_n \xrightarrow{d} \xi$, prove $\sigma_n^2 \to \sigma$.

7. Let $\{\xi_n\}_{n\geqslant 1}$ be i.i.d. random variables with $\mathbb{P}(\xi_i = 0) = \mathbb{P}(\xi_i = 1) = 1/2$. Calculate the distribution and characteristic function of $\xi = 2\sum_{j=1}^{\infty} \xi_j/3^j$.

8. Exemplify that vague convergence is not equivalent to weak convergence.

9. Let $\{\mu_n, \mu\}_{n\geqslant 1}$ be finite measures on a metric space E. Prove the following:
   (a) $\mu_n \xrightarrow{w} \mu$ if and only if $\mu_k(A) \to \mu(A)$ for any μ-continuous compact A.
   (b) $\mu_k \xrightarrow{w} \mu$ if and only if $\mu_k(A) \to \mu(A)$ for any μ-continuous open A.
   (c) When $E = \mathbb{R}^n$, $\mu_k \xrightarrow{v} \mu$ if and only if $\mu_k(I) \to \mu(I)$ for any finite μ-continuous interval I.

10. Let $g \geqslant 0$ be a continuous function on $\mathbb{R}^n$ and $\{\xi_k, \xi\}_{k\geqslant 1}$ are n-dimensional random variables. If $\xi_n \xrightarrow{w} \xi$, prove

$$\varliminf_{n\to\infty} \mathbb{E}g(\xi_n) \geqslant \mathbb{E}g(\xi).$$

11. Let $\{F_n, F\}_{k\geqslant 1}$ be probability distribution functions on $\mathbb{R}^n$ such that is continuous. Prove that $F_n \to F$ implies $\sup_x |F_n(x) - F(x)| \to 0$.

12. Let $\{\mu_n, \mu\}_{n\geqslant 1}$ be finite measures on a measurable space $(E, \mathscr{E})$. Prove

$$\sup_{A\in\mathscr{E}} |\mu_n(A) - \mu(A)| \to 0, \quad n \uparrow \infty$$

is equivalent to

$$\sup_{f\in\mathscr{E},\ |f|\leqslant 1} |\mu_n(f) - \mu(f)| \to 0, \quad n \uparrow \infty.$$

13. Prove that a family of probability measures $\{\mu_t, t \in T\}$ on $\mathbb{R}^n$ is tight if and only if there exists an increasing function $\phi : \mathbb{R}^+ \to \mathbb{R}^+$ with $\lim_{x\to\infty} \phi(x) = \infty$ such that $\sup_{t\in T} \mu_t\big(\phi(|\cdot|)\big) < \infty$.

14. Let $\{\xi_k\}_{k\geqslant 1}$ be a sequence of random variables on $\mathbb{R}^n$ such that $\{\mathbb{P}\circ\xi_k^{-1}\}_{k\geqslant 1}$ is tight. Prove that for any random variables $\eta_n \xrightarrow{\mathbb{P}} 0$, there holds $\xi_n\eta_n \xrightarrow{\mathbb{P}} 0$.

15. Let $h : \mathbb{R} \to \mathbb{R}$ be measurable, and let D_h be the class of discontinuous points of h. Prove that if D_h is measurable, then for any finite measures $\{\mu_n, \mu\}_{n\geqslant 1}$ on $\mathbb{R}$ such that $\mu_n \xrightarrow{w} \mu$ and $\mu(D_h) = 0$, then $\mu_n \circ h^{-1} \xrightarrow{w} \mu \circ h^{-1}$.

16. Let $\mu(\mathrm{d}x) = p(x)\,\mathrm{d}x$ be a finite measure on $(\mathbb{R}, \mathscr{B})$.
    (a) Prove $\lim_{|t|\to\infty} f_\mu(t) = 0$ (*Hint*: $\lim_{|t|\to\infty} \frac{\int_0^t f_\mu(s)}{t} = 0$).
    (b) If p has integrable derivative function p', then $\lim_{|t|\to\infty} t f_\mu(t) = 0$.
    (c) What happens if p has integrable derivatives $p^{(k)}$ for $1 \leqslant k \leqslant n$ for some $n \geqslant 2$?

17. Let μ be a finite measure on $\mathbb{R}$. Prove that for any $x \in \mathbb{R}$,

$$\mu(\{x\}) = \lim_{T\to\infty} \frac{1}{2T} \int_{-T}^{T} \mathrm{e}^{-\mathrm{i}\,tx} f_\mu(t)\,\mathrm{d}t.$$

Chapter 7

Probability Distances

Let (E, ρ) be a metric space with Borel σ-algebra $\mathscr{E}$, and let $\mathscr{P}(E)$ be class of all probability measures on $(E, \mathscr{E})$. In this chapter, we introduce some distances on $\mathscr{P}(E)$, including the metrization of weak topology, the total variation distance for the uniform convergence, and Wasserstein distance arising from optimal transport.

7.1 Metrization of Weak Topology

Let (E, ρ) be a Polish space. Then space $C_b(E)$ of bounded continuous functions is also a Polish space under uniform norm $\|f\|_\infty = \sup_E |f|$ (see [18–20]). For a dense sequence $\{f_n\}_{n\geqslant 1}$ in $C_b(E)$, define

$$d_w(\mu, \nu) := \sum_{n=1}^{\infty} 2^{-n} \{|\mu(f_n) - \nu(f_n)| \wedge 1\}, \quad \mu, \quad \nu \in \mathscr{P}(E).$$

Theorem 7.1. *Let (E, ρ) be a Polish space. Then $(\mathscr{P}(E), d_w)$ is a separable metric space, and for any $\{\mu_n\}_{n\geqslant 1} \subset \mathscr{P}(E)$ and $\mu \in \mathscr{P}(E)$, $\mu_n \xrightarrow{w} \mu$ if and only if $d_w(\mu_n, \mu) \to 0$. If E is compact, then $(\mathscr{P}(E), d_w)$ is complete.*

Proof. (a) $\boldsymbol{d_w}$ *is a distance*: Obviously, $d_w(\mu, \mu) = 0$. If $d_w(\mu, \nu) = 0$, then $\mu(f_n) - \nu(f_n) = 0 (\forall n)$. Since $\{f_n\}_{n\geqslant 1}$ is dense in $C_b(E)$, it follows that $\mu(f) = \nu(f)$ for any $f \in C_b(E)$, thus $\mu = \nu$ by Lemma 6.9. Finally, d_w clearly satisfies the triangle inequality.

(b) *Equivalence to the weak topology*: Obviously, if $\mu_n \xrightarrow{w} \mu$, then $d_w(\mu_n, \mu) \to 0$. Conversely, let $d_w(\mu_n, \mu) \to 0$. We are going to prove $\mu_n(f) - \mu(f) \to 0$ for any $f \in C_b(E)$. Given $f \in C_b(E)$, since $\{f_n\}$ is dense in $C_b(E)$, for any $\varepsilon > 0$, there exists $n_0 \geqslant 1$ such that $||f_{n_0} - f||_\infty < \varepsilon$. So,

$$\begin{aligned}\overline{\lim_{n\to\infty}} |\mu_n(f) - \mu(f)| &\leqslant 2\varepsilon + \overline{\lim_{n\to\infty}} |\mu_n(f_{n_0}) - \mu(f_{n_0})| \\ &\leqslant 2\varepsilon + 2^{n_0+1} \overline{\lim_{n\to\infty}} d_w(\mu_n, \mu) \\ &= 2\varepsilon.\end{aligned}$$

As ε is arbitrary, we have $\mu_n(f) \to \mu(f)$.

(c) *Separability*: $\forall m \geqslant 1$, let $U_m = \{(\mu(f_1), \ldots, \mu(f_m)) : \mu \in \mathscr{P}(E)\} \subset \mathbb{R}^m$. Since $\mathbb{R}^m$ is separable, so is U_m. Thus, there exists a countable set $\mathscr{P}_m \subset \mathscr{P}(E)$ such that

$$\tilde{U}_m := \{(\mu(f_1), \ldots, \mu(f_m)) : \mu \in \mathscr{P}_m\}$$

is dense in U_m. Thus, $\mathscr{P}_\infty := \bigcup_{m=1}^{\infty} \mathscr{P}_m$ is a countable subset of $\mathscr{P}(E)$, so that it suffices to prove that $\mathscr{P}_\infty$ is dense in $\mathscr{P}(E)$ under distance d_w.

In fact, for any $\mu \in \mathscr{P}(E)$, there exists $\mu_m \in \mathscr{P}_m$ such that

$$|\mu_m(f_i) - \mu(f_i)| \leqslant \frac{1}{m}, \quad \forall 1 \leqslant i \leqslant m.$$

Thus,

$$d_w(\mu_m, \mu) \leqslant 2^{-m} + \frac{1}{m} \to 0 \; (m \to \infty).$$

(d) *Completeness of* d_w: Assume E is locally compact. Note $\{\mu_n\}_{n\geqslant 1} \subset \mathscr{P}(E)$ is a Cauchy sequence under d_w. Then $\forall m \geqslant 1, \{\mu_n(f_m)\}_{n\geqslant 1}$ is a Cauchy sequence, so converge to some number, denoted by $\phi(f_m)$. Moreover, given $f \in C_b(E), \forall \varepsilon > 0, \exists m_0 \geqslant 1$ such that $||f_{m_0} - f||_\infty < \varepsilon$. Thus,

$$\begin{aligned}\overline{\lim_{m,n\to\infty}} |\mu_m(f) - \mu_n(f)| &\leqslant 2\varepsilon + \overline{\lim_{m,n\to\infty}} |\mu_m(f_{m_0}) - \mu_n(f_{m_0})| \\ &= 2\varepsilon.\end{aligned}$$

As ε is arbitrary, we note that $\{\mu_n(f)\}_{n\geqslant 1}$ is also a Cauchy sequence, which converge to some number, denoted by $\phi(f)$. By the properties of integral, it follows that

$$\phi : C_b(E) \to \mathbb{R}$$

is a linear map, $\phi(1) = 1$, and $\phi(f) \geqslant 0$ for $f \geqslant 0$. By Riesz's representation theorem, there exists unique $\mu \in \mathscr{P}(E)$ such that $\mu(f) = \phi(f)$ for every $f \in C_b(E)$, see [11, Theorem IV.14]. By the construction of ϕ, it follows that $\mu_n \xrightarrow{w} \mu$, hence $d_w(\mu_n, \mu) \to 0$ from (b). □

7.2 Wasserstein Distance and Optimal Transport

In this section, we introduce the transportation problem initiated by Monge in 1781 and characterized by Kantorovich in the 1940s using couplings, which leads to the notion of the Wasserstein distance. In particular, when E is a Polish space, $\mathscr{P}(E)$ is also a Polish space under the Wasserstein distance.

Definition 7.2. Let $\mathscr{P}(E)$ be the class of probability measures on a measurable space $(E, \mathscr{E})$, and let $\mu, \nu \in \mathscr{P}(E)$. A probability measure π on the product space $(E \times E, \mathscr{E} \times \mathscr{E})$ is called a coupling of μ and ν, denoted by $\pi \in \mathscr{C}(\mu, \nu)$, if its marginals are μ and ν, i.e.

$$\pi(A \times E) = \mu(A), \quad \pi(E \times A) = \nu(A), \quad A \in \mathscr{E}.$$

A simple coupling is the product measure $\mu \times \nu$, which is called the independent coupling of μ and ν. Therefore, $\mathscr{C}(\mu, \nu) \neq \varnothing$.

7.2.1 Transport problem, coupling and Wasserstein distance

Before introducing the general theory, let us consider a simple example.

Let $x_1, \ldots, x_n$ be n many cities, each city produces and consumes certain product. Let μ and ν be the produced (initial) distribution and the consumed (target) distribution, respectively. We intend to design a scheme to transport the product from initial distribution μ to the target distribution ν.

Let $\mu(\{x_i\}) = \mu_i,\ \nu(\{x_i\}) = \nu_i,\ 1 \leqslant i \leqslant n$. We have $\mu_i, \nu_i \geqslant 0$ and $\sum_{i=1}^n \mu_i = \sum_{i=1}^n \nu_i = 1$. So, μ and ν are probability measures on space $E := \{x_1, \ldots, x_n\}$. Let $\pi = \{\pi_{ij} : 1 \leqslant i, j \leqslant n\}$ be a transport scheme, where $\pi_{ij} \geqslant 0$ denotes the amount of product transported from x_i to x_j. Then, the scheme π transports μ into ν if and only if

$$\mu_i = \sum_{j=1}^n \pi_{ij},\quad \nu_i = \sum_{j=1}^n \pi_{ji},\quad 1 \leqslant i \leqslant n.$$

Thus, π is a scheme π transporting μ into ν if and only if $\pi \in \mathscr{C}(\mu, \nu)$.

Let $\rho_{ij} \geqslant 0$ be the cost to transport a unit product from x_i to x_j, which is called the cost function. Then for any scheme $\pi \in \mathscr{C}(\mu, \nu)$, the total cost is

$$\sum_{i,j=1}^n \rho_{ij}\pi_{ij} = \int_{E\times E} \rho \, \mathrm{d}\pi.$$

Thus, the lowest cost to transport from μ to ν is

$$W_1^\rho(\mu, \nu) := \inf_{\pi\in\mathscr{C}(\mu,\nu)} \int_{E\times E} \rho \, \mathrm{d}\pi,$$

which is called L^1-Wasserstein distance between μ and ν induced by ρ.

In general, we define the L^p-Wasserstein distance on $\mathscr{P}(E)$ over a metric space $(E, \mathscr{E})$ as follows.

Definition 7.3. Let (E, ρ) be a metric space. $\forall p \in [1, \infty)$, define the L^p-Wasserstein distance induced by ρ as

$$W_p(\mu, \nu) := \inf_{\pi\in\mathscr{C}(\mu,\nu)} \left\{\int_{E\times E} \rho^p \, \mathrm{d}\pi\right\}^{1/p},\quad \mu,\ \nu \in \mathscr{P}(E).$$

A coupling π is called optimal if it reaches the infimum.

In generally, ρ may be unbounded, so that $W_p(\mu, \nu)$ may be infinite for some $\mu, \nu \in \mathscr{P}(E)$. To make W_p finite, we restrict to the following subspace of $\mathscr{P}(E)$ of finite p-moment:

$$\mathscr{P}_p(E) = \{\mu \in \mathscr{P}(E) : \mu(\rho(o, \cdot)) < \infty\},\quad p \geqslant 1,$$

where $o \in E$ is any fixed point. By the triangle inequality, the definition of $\mathscr{P}_p(E)$ is independent of the choice of $o \in E$.

7.2.2 Optimal coupling and Kantorovich's dual formula

We first consider the existence of optimal coupling.

Theorem 7.4. *Let (E, ρ) be a Polish space. Then $\forall \mu, \nu \in \mathscr{P}_p(E)$, there exists $\pi \in \mathscr{C}(\mu, \nu)$ such that $W_p(\mu, \nu) = \pi(\rho^p)$.*

Proof. Since $\mu, \nu \in \mathscr{P}_p(E)$ and $\mu \times \nu \in \mathscr{C}(\mu, \nu)$, we have

$$\begin{aligned} W_p(\mu, \nu)^p &\leqslant \int_{E\times E} \rho^p(x, y)\mu(\mathrm{d}x)\nu(\mathrm{d}y) \\ &\leqslant 2^{p-1} \int_{E\times E} (\rho^p(x, o) + \rho^p(y, o))\mu(\mathrm{d}x)\nu(\mathrm{d}y) \\ &< \infty. \end{aligned}$$

Thus, for any $n \geqslant 1$, there exists $\pi_n \in \mathscr{C}(\mu, \nu)$ such that

$$W_p(\mu, \nu)^p \geqslant \pi_n(\rho^p) - \frac{1}{n}. \tag{7.1}$$

So, if π_n converge weakly to some π_0, then π_0 should be an optimal coupling. For this, we first prove that $\{\pi_n\}_{n\geqslant 1}$ is tight. In fact, by Theorem 6.16, we know that finite set $\{\mu, \nu\}$ is tight, so for any $\varepsilon > 0$, there exists a compact set $K \subset E$ such that $\mu(K^c) + \nu(K^c) < \varepsilon$. Thus, $\forall \pi \in \mathscr{C}(\mu, \nu), \pi((K \times K)^c) \leqslant \pi(K^c \times E) + \pi(E \times K^c) = \mu(K^c) + \nu(K^c) < \varepsilon$. Therefore, $\mathscr{C}(\mu, \nu)$ is tight. Hence, there exist a subsequence $\{\pi_{n_k}\}_{k\geqslant 1}$ and $\pi_0 \in \mathscr{P}(E)$ such that $\pi_{n_k} \xrightarrow{w} \pi_0$ $(k \to \infty)$. Obviously, $\pi_0 \in \mathscr{C}(\mu, \nu)$. Combining this with (7.1), we obtain that for any $N \in (0, \infty)$,

$$\pi_0(\rho^p \wedge N) = \lim_{k\to\infty} \pi_{n_k}(\rho^p \wedge N) \leqslant W_p(\mu, \nu)^p.$$

Letting $N \uparrow \infty$ gives $\pi_0(\rho^p) \leqslant W_p(\mu, \nu)^p$. □

From Definition 7.3, it is easy to derive an upper bound estimate on the Wasserstein distance. To estimate it from below, we introduce the Kantorovich dual formula by using the following classes of function pairs for $\mu, \nu \in \mathscr{P}(E)$:

$$\mathscr{F}_{\mu,\nu} = \{(f, g) : f \in L^1(\mu), g \in L^1(\nu), f(x) \leqslant g(y) + \rho(x, y)^p, \forall x, y \in E\},$$

$$\mathscr{F}_{\text{Lip}} = \{(f, g) : f, g \text{ Lipschitz continuous } f(x) \leqslant g(y) + \rho(x, y)^p, x, y \in E\}.$$

Theorem 7.5 (Kantorovich's dual formula). *Let (E, ρ) be a Polish space. Then $\forall \mu, \nu \in \mathscr{P}_p(E)$,*

$$W_p(\mu, \nu)^p = \sup_{(f,g)\in\mathscr{F}_{\mu,\nu}} \{\mu(f) - \nu(g)\} = \sup_{(f,g)\in\mathscr{F}_{\mathrm{Lip}}} \{\mu(f) - \nu(g)\}. \tag{7.2}$$

Proof. Since $\mathscr{F}_{\mathrm{Lip}} \subset \mathscr{F}_{\mu,\nu}$, we only need to prove

$$\sup_{(f,g)\in\mathscr{F}_{\mu,\nu}} \{\mu(f) - \nu(g)\} \leqslant W_p(\mu, \nu)^p \leqslant \sup_{(f,g)\in\mathscr{F}_{\mathrm{Lip}}} \{\mu(f) - \nu(g)\}.$$

In the following, we only prove the first inequality, as the second is far from trivial, see, for instance, [10, Section 3] or [3, Chapter 5] for details.

Let $(f, g) \in \mathscr{F}_{\mu,\nu}$ and $\pi \in \mathscr{C}(\mu, \nu)$. We have

$$\mu(f)-\nu(g) = \int_{E\times E} (f(x)-g(y))\pi(\mathrm{d}x, \mathrm{d}y) \leqslant \int_{E\times E} \rho(x, y)^p \pi(\mathrm{d}x, \mathrm{d}y).$$

Thus, the first equation follows from the definition of W_p. □

7.2.3 The metric space $(\mathscr{P}_p(E), W_p)$

Theorem 7.6. *Let (E, ρ) be a Polish space. Then $(\mathscr{P}_p(E), W_p)$ is also a Polish space.*

Proof. (a) First, we prove W_p is a metric. Obviously, $W_p(\mu, \nu) = 0$ if and only if $\mu = \nu$, so we only need to prove the triangle inequality.

$\forall \mu_1, \mu_2, \mu_3 \in \mathscr{P}_p(E)$, let π_{12} and π_{23} be optimal couplings of (μ_1, μ_2) and (μ_2, μ_3), respectively. We have

$$W_p(\mu_1, \mu_2) = \pi_{12}(\rho^p)^{1/p}, \ W_p(\mu_2, \mu_3) = \pi_{23}(\rho^p)^{1/p}.$$

To construct the optimal coupling of μ_1 and μ_3, let $\pi_{12}(x_1, \mathrm{d}x_2)$ be the regular conditional probability of π_{12} for given x_1 and $\pi_{23}(x_2, \mathrm{d}x_3)$ be the regular conditional probability of π_{23}

for given x_2. Set

$$\pi_{13}(A \times B) = \mu_1(A) \int_E \pi_{23}(x_2, B)\pi_{12}(x_1, \mathrm{d}x_2).$$

It is clear that $\pi_{13} \in \mathscr{C}(\mu_1, \mu_3)$. Then

$$\pi(\mathrm{d}x_1, \mathrm{d}x_2, \mathrm{d}x_3) := \mu_1(\mathrm{d}x_1)\pi_{12}(x_1, \mathrm{d}x_2)\pi_{23}(x_2, \mathrm{d}x_3)$$

is a probability measure on $E \times E \times E$, and for

$$\rho_{ij}(x_1, x_2, x_3) := \rho(x_i, x_j),\ 1 \leqslant i, j \leqslant 3,$$

we have

$$\pi(\rho_{ij}^p) = \pi_{ij}(\rho^p),\ 1 \leqslant i, j \leqslant 3.$$

Thus, by the triangle inequality in $L^p(\pi)$,

$$\begin{aligned} W_p(\mu_1, \mu_3) &\leqslant \pi(\rho_{13}^p)^{1/p} \leqslant \pi((\rho_{12} + \rho_{23})^p)^{1/p} \\ &\leqslant \pi(\rho_{12}^p)^{1/p} + \pi(\rho_{23}^p)^{1/p} \\ &= W_p(\mu_1, \mu_2) + W_p(\mu_2, \mu_3). \end{aligned}$$

(b) Next, we prove W_p is complete. Let $\{\mu_n\}_{n\geqslant 1} \subset \mathscr{P}_p(E)$ be a Cauchy sequence under W_p. Then $\{\mu_n\}_{n\geqslant 1} \subset \mathscr{P}_p(E)$ is tight (see Lemma 6.14 in [13]). Without loss of generality, we assume that $\mu_n \xrightarrow{w} \mu$ for some $\mu \in \mathscr{P}(E)$. On the other hand, given $o \in E$, we have

$$\mu_n(\rho(o, \cdot)^p) \leqslant 2^{p-1}\mu_1(\rho(o, \cdot)^p) + 2^{p-1}W_p(\mu_1, \mu_n)^p,$$

which are bounded for $n \geqslant 1$, so $\exists C > 0$ such that $\forall N \geqslant 1$

$$\mu(\rho(o, \cdot)^p \wedge N) = \lim_{n\to\infty} \mu_n(\rho(o, \cdot)^p \wedge N) \leqslant C.$$

Thus, $\mu \in \mathscr{P}_p(E)$ and

$$\varliminf_{n\to\infty} \mu_n(\rho(o, \cdot)^p) \geqslant \mu(\rho(o, \cdot)^p). \tag{7.3}$$

Moreover, $\forall \varepsilon > 0, \exists n_0 \geqslant 1$ such that $W_p(\mu_{n_0}, \mu_n)^p \leqslant \varepsilon, \forall n \geqslant n_0$. Then

$$
\begin{aligned}
&\mu_n((N - \rho(o,\cdot)^p)^+) \\
&\leqslant \mu_{n_0}((N - \rho(o,\cdot)^p)^+) + |\mu_n((N - \rho(o,\cdot)^p)^+) - \mu_{n_0}((N - \rho(o,\cdot)^p)^+)| \\
&\leqslant \mu_{n_0}((N - \rho(o,\cdot)^p)^+) + 2^{p-1} W_p(\mu_n, \mu_{n_0})^p \\
&\leqslant \mu_{n_0}((N - \rho(o,\cdot)^p)^+) + 2^{p-1}\varepsilon.
\end{aligned}
$$

Hence,

$$\overline{\lim_{n\to\infty}} \mu_n(\rho(o,\cdot)^p) \leqslant \overline{\lim_{n\to\infty}} \mu_n(\rho(o,\cdot)^p \wedge N) + 2^{p-1}\varepsilon.$$

As ε is arbitrary and $\mu_n \xrightarrow{w} \mu$, we have

$$\overline{\lim_{n\to\infty}} \mu_n(\rho(o,\cdot)^p) \leqslant \overline{\lim_{n\to\infty}} \mu_n(\rho(o,\cdot)^p \wedge N) = \mu(\rho(o,\cdot)^p \wedge N) \leqslant \mu(\rho(o,\cdot)^p).$$

From this and (7.3), it follows that $\mu(\rho(o,\cdot)^p) = \lim_{n\to\infty} \mu_n(\rho(o,\cdot)^p)$. Thus, by Kantorovich's dual formula, the dominated convergence theorem, and $W_p(\mu_n, \mu_m) \to 0$ as $n, m \to \infty$, we obtain

$$
\begin{aligned}
\lim_{n\to\infty} W_p(\mu, \mu_n)^p &= \lim_{n\to\infty} \sup_{(f,g)\in\mathscr{F}_{Lip}} |\mu(f) - \mu_n(g)| \\
&= \lim_{n\to\infty} \sup_{(f,g)\in\mathscr{F}_{Lip}} \lim_{m\to\infty} |\mu_m(f) - \mu_n(g)| W_p(\mu, \mu_n)^p \\
&\leqslant \lim_{n\to\infty} \lim_{m\to\infty} W_p(\mu_m, \mu_n)^p = 0.
\end{aligned}
$$

(c) Finally, we prove W_p is separable. $\forall N \geqslant 1$, let

$$\mathscr{P}_p^{(N)}(E) = \left\{\mu \in \mathscr{P}_p(E) : \text{supp}\mu \subset \bar{B}(o, N)\right\},$$

where $\bar{B}(o,N)$ is a closed ball with radium N centered at o. As $\forall \mu \in \mathscr{P}_p(E)$, it is easy to prove $N \to \infty$,

$$\mu_N := \frac{\mu(\cdot \cap B(o,N))}{\mu(B(o,N))} \xrightarrow{W_p} \mu,$$

so we have $\bigcup_{N=1}^{\infty} \mathscr{P}_p^{(N)}(E)$ is dense in $(\mathscr{P}_p(E), W_p)$. Thus, we only need to prove that each $\mathscr{P}_p^{(N)}(E)$ is separable. Since $\rho(o,\cdot)$ is bounded on $\bar{B}(o,N)$, as shown in step (b), the weak convergence is equivalent to the convergence in W_p (see Exercise 6). Then the proof is finished by Theorem 7.1, which says that $\mathscr{P}_p^{(N)}(E)$ is separable under weak topology. □

Theorem 7.7. *Let $\mathfrak{M} \subset \mathscr{P}_p(E)$. Then $\mathfrak{M}$ is compact under W_p if and only if it is weakly compact and*

$$\lim_{N\to\infty} \sup_{\mu\in\mathfrak{M}} \mu\left(\rho(o,\cdot)^p \mathbf{1}_{\{\rho(o,\cdot)\geqslant N\}}\right) = 0. \tag{7.4}$$

Proof. (a) Necessity. It is clear that $W_p(\mu_n,\mu) \to 0$ implies $\mu_n(f) \to \mu(f)$ for any Lipschitz continuous function f, so the topology induced by W_p is stronger that weak topology. Thus, if $\mathfrak{M}$ is compact under W_p, then $\mathfrak{M}$ is also compact under weak topology. It remains to prove (7.4).

Since $\mathfrak{M}$ is compact under W_p, for any $\varepsilon > 0$, there exist $\mu_1,\ldots,\mu_n \in \mathfrak{M}$ such that

$$\min_{1\leqslant i\leqslant n} W_p(\mu_i,\mu)^p < \varepsilon, \quad \mu\in\mathfrak{M}.$$

Thus,

$$\begin{aligned}\mu\left((\rho(o,\cdot)^p - N)^+\right) &\leqslant \max_{1\leqslant i\leqslant n} \mu_i\left((\rho(o,\cdot)^p - N)^+\right) + 2^{p-1}W_p(\mu_i,\mu)^p \\ &\leqslant \sum_{i=1}^n \mu_i\left((\rho(o,\cdot)^p - N)^+\right) + 2^{p-1}\varepsilon, \quad \mu\in\mathfrak{M}.\end{aligned}$$

Hence,

$$\begin{aligned}\overline{\lim_{N\to\infty}} \sup_{\mu\in\mathfrak{M}} \mu\left(\rho(o,\cdot)^p \mathbf{1}_{\{\rho(o,\cdot)\geqslant N\}}\right) &\leqslant 2\, \overline{\lim_{N\to\infty}} \sup_{\mu\in\mathfrak{M}} \mu\left(\left(\rho(o,\cdot)^p - \frac{N}{2}\right)^+\right) \\ &\leqslant 2^p\varepsilon.\end{aligned}$$

As ε is arbitrary, we get (7.4) immediately.

(b) Sufficiency. Let $\mathfrak{M}$ be weakly compact and (7.4) hold. We intend to prove that $\mathfrak{M}$ is compact under W_p. For this, we only need to prove that for any sequence $\{\mu_n\}_{n\geqslant 1} \subset \mathfrak{M}$, there exists a convergent subsequence under W_p. By the weak compactness of $\mathfrak{M}$, we may and do assume that $\mu_n \xrightarrow{w} \mu$. Let $\{x_1, x_2, \ldots\}$ be a dense subset of E. Then we have $\bigcup_{i=1}^{\infty} B(x_i, \varepsilon) \supset E$ for any $\varepsilon > 0$, where $B(x_i, \varepsilon)$ is an open ball with radium ε centered at x_i. Since set $\{\varepsilon > 0 : \exists i \geqslant 1 \text{ such that } \mu(\partial B(x_i, \varepsilon)) > 0\}$ is at most countable, for $m \geqslant 1$, take $\varepsilon_m \in (0, 1/m)$ such that $B(x_i, \varepsilon_m)$ are all μ continuous sets. Let

$$U_1 = B(x_1, \varepsilon_m), U_{i+1} = B(x_{i+1}, \varepsilon_m) \setminus \bigcup_{j=1}^{i} B(x_j, \varepsilon_m).$$

Then $\{U_i\}_{i\geqslant 1}$ is a sequence of mutually disjoint μ-continuous sets, $\sum_{i=1}^{\infty} U_i = E$ and the radium of U_i is less than $\frac{1}{m}$. Let $r_n = \sum_{i=1}^{\infty} \mu_n(U_i) \wedge \mu(U_i)$. Then $r_n \in [0, 1]$ with $\lim_{n\to\infty} r_n = 1$. Let

$$Q_n(\mathrm{d}x) = \mu_n(\mathrm{d}x) - \sum_{i=1}^{\infty} \frac{\mu_n(U_i) \wedge \mu(U_i)}{\mu_n(U_i)} \mathbf{1}_{U_i}(x)\mu_n(\mathrm{d}x),$$

$$Q(\mathrm{d}x) = \mu(\mathrm{d}x) - \sum_{i=1}^{\infty} \frac{\mu_n(U_i) \wedge \mu(U_i)}{\mu(U_i)} \mathbf{1}_{U_i}(x)\mu(\mathrm{d}x).$$

Then

$$\pi_n(\mathrm{d}x, \mathrm{d}y) := \sum_{i=1}^{\infty} \mathbf{1}_{U_i}(x)\mathbf{1}_{U_i}(y) \frac{\mu_n(U_i) \wedge \mu(U_i)}{\mu_n(U_i)\mu(U_i)} \mu_n(\mathrm{d}x)\mu(\mathrm{d}y) + \frac{1}{1 - r_n} Q_n(\mathrm{d}x)Q(\mathrm{d}y)$$

is a coupling of μ_n and μ (if $r_n = 1$, then the last term is set to 0), hence

$$\begin{aligned} W_p(\mu_n, \mu)^p &\leqslant \pi_n(\rho^p) \leqslant m^{-p} + \frac{2^{p-1}}{1 - r_n} \left(Q_n(\rho(o, \cdot)^p) + Q(\rho(o, \cdot)^p)\right) \\ &\leqslant m^{-p} + 2^p N^p (1 - r_n) + 2^{p-1} \sup_{k\geqslant 1} \mu_k \left(\rho(o, \cdot)^p \mathbf{1}_{\{\rho(o,\cdot)\geqslant N\}}\right) \\ &\quad + 2^{p-1} \mu \left(\rho(o, \cdot)^p \mathbf{1}_{\{\rho(o,\cdot)\geqslant N\}}\right). \end{aligned}$$

By first letting $n \to \infty$, then $N \to \infty$ and finally $m \to \infty$, we obtain $W_p(\mu_n, \mu) \to 0 (n \to \infty)$. □

7.3 Total Variation Distance

Let $\mathscr{P}(E)$ be the class of probability measures on a measurable space $(E, \mathscr{E})$. The total variation distance on $\mathscr{P}(E)$ is defined as

$$||\mu - \nu||_{\text{Var}} := |\mu - \nu|(E) = 2(\mu - \nu)^+(E) = 2(\nu - \mu)^+(E). \quad (7.5)$$

We will characterize this distance by using the Wasserstein coupling, we define the wedge $\mu \wedge \nu$ of μ and ν.

Proposition 7.8. *For any $\mu, \nu \in \mathscr{P}(E)$,*

$$\mu \wedge \nu := \mu - (\mu - \nu)^+ = \nu - (\nu - \mu)^+$$

is a sub-probability measure, i.e. it is a measure with $\mu \wedge \nu \leqslant 1$.

Proof. Since $\mu \geqslant (\mu - \nu)^+$ and $\nu \geqslant (\nu - \mu)^+$, both $\mu - (\mu - \nu)^+$ and $\nu - (\nu - \mu)^+$ are sub-probability measures. It suffices to prove that they are equal. By Hahn's decomposition theorem, there exists $D \in \mathscr{E}$ such that $(\mu - \nu)(D) = \inf_{A \in \mathscr{E}} (\mu - \nu)(A)$, and for any $A \in \mathscr{E}$,

$$(\mu - \nu)^+(A) = (\mu - \nu)(D^c \cap A), \quad (\nu - \mu)^+(A) = (\nu - \mu)(A \cap D).$$

Thus,

$$\begin{aligned}(\mu - (\mu - \nu)^+)(A) &= \mu(A) - \mu(D^c \cap A) + \nu(D^c \cap A) \\ &= \mu(A \cap D) + \nu(A) - \nu(D \cap A) \\ &= (\nu - (\nu - \mu)^+)(A).\end{aligned}$$ □

The Wasserstein coupling of μ and ν is defined as

$$\pi_0(\mathrm{d}x, \mathrm{d}y) := (\mu \wedge \nu)(\mathrm{d}x)\delta_x(\mathrm{d}y) + \frac{(\mu - \nu)^+(\mathrm{d}x)(\mu - \nu)^-(\mathrm{d}y)}{(\mu - \nu)^-(E)},$$

where for $\mu = \nu$, we set $\frac{(\mu-\nu)^+(\mathrm{d}x)(\mu-\nu)^-(\mathrm{d}y)}{(\mu-\nu)^-(E)} = 0$.

Regarding a coupling as a scheme to transport μ into ν, the idea of this coupling is that to keep the common part of μ and ν without transport and to transport $(\mu-\nu)^+$ to $(\nu-\mu)^+$ using the independent coupling. The following result shows that Wasserstein coupling gives an optimal transport under the cost function $\mathbf{1}_{\{x\neq y\}}$.

Theorem 7.9. *Let* $D_0 = \{(x,x) : x \in E\} \in \mathscr{E} \times \mathscr{E}$. *Then* $\pi_0(\mathrm{d}x, \mathrm{d}y) \in \mathscr{C}(\mu,\nu)$, *and*

$$\|\mu-\nu\|_{\mathrm{Var}} = 2 \inf_{\pi\in\mathscr{C}(\mu,\nu)} \pi(D_0^c) = 2\pi_0(D_0^c). \tag{7.6}$$

Proof. (a) Obviously, π_0 is a probability measure on the product space $(E \times E, \mathscr{E} \times \mathscr{E})$. When $\mu = \nu$, we have $\pi_0(\mathrm{d}x, \mathrm{d}y) = \mu(\mathrm{d}x)\delta_x(\mathrm{d}y)$, so that

$$\pi_0(A \times E) = \pi_0(E \times A) = \mu(A), \quad A \in \mathscr{E}.$$

Hence, $\pi_0 \in \mathscr{C}(\mu,\nu)$.

When $\mu \neq \nu$, we have $(\mu-\nu)^+(E) > 0$. Since $(\mu-\nu)^- = (\nu-\mu)^+$ and $\mu(E) = \nu(E) = 1$, we have $(\mu-\nu)^-(E) = (\mu-\nu)^+(E)$. Thus, for any $A \in \mathscr{E}$,

$$\begin{aligned}\pi_0(A \times E) &= (\mu\wedge\nu)(A) + \frac{(\mu-\nu)^+(A)(\mu-\nu)^-(E)}{(\mu-\nu)^-(E)}\\ &= \mu(A) - (\mu-\nu)^+(A) + (\mu-\nu)^+(A) = \mu(A),\\ \pi_0(E \times A) &= \int_E \mathbf{1}_A(x)(\mu\wedge\nu)(\mathrm{d}x) + (\mu-\nu)^-(A)\\ &= \nu(A) - (\mu-\nu)^-(A) + (\mu-\nu)^-(A)\\ &= \nu(A).\end{aligned}$$

Hence, $\pi_0 \in \mathscr{C}(\mu,\nu)$.

(b) $\forall \pi \in \mathscr{C}(\mu,\nu)$, we have

$$\begin{aligned}\mu(A) - \nu(A) &= \pi(A \times E) - \pi(E \times A)\\ &\leqslant \pi(\{(x,y) : x \in A, y \notin A\})\\ &\leqslant \pi(D_0^c).\end{aligned}$$

Thus, $\|\mu-\nu\|_{\mathrm{Var}} \leqslant 2\pi(D_0^c)$. To prove equation (7.6), we only need to prove $\|\mu-\nu\|_{\mathrm{Var}} \geqslant 2\pi_0(D_0^c)$. We prove in the case of $\mu \neq \nu$.

By (7.5) and the definition of π_0, it follows

$$\begin{aligned}\pi_0(D_0^c) &= \frac{1}{(\mu-\nu)^+(E)}\int_{D_0^c}(\mu-\nu)^+(\mathrm{d}x)(\mu-\nu)^-(\mathrm{d}y)\\ &\leqslant \frac{1}{(\mu-\nu)^+(E)}\int_{E\times E}(\mu-\nu)^+(\mathrm{d}x)(\mu-\nu)^-(\mathrm{d}y)\\ &= (\mu-\nu)^-(E) = \frac{1}{2}\|\mu-\nu\|_{\mathrm{Var}}.\end{aligned}$$

□

7.4 Exercises

1. Let (E,ρ) be a Polish space. On $\mathscr{P}(E)$, construct a metric equivalent to the vague convergence and give a proof. Is this metric complete?

2. Let $(E,\mathscr{E})$ be a measurable space. Prove that $\forall \mu,\nu\in\mathscr{P}(E)$,

$$\|\mu-\nu\|_{\mathrm{Var}} = 2\sup_{A\in\mathscr{E}}|\mu(A)-\nu(A)| = |\mu-\nu|(E).$$

3. Let $V\geqslant 1$ be a measurable function on a measurable space $(E,\mathscr{E})$. Prove that $\forall\mu\in\mathscr{P}(E)$, define weighted variance

$$\|\mu\|_V = \int_E V(x)\mu(\mathrm{d}x).$$

Prove for measurable function f,

$$\sup_{\|\mu\|_V\leqslant 1}\int_E |f|\,\mathrm{d}\mu = \sup_{x\in E}\frac{|f(x)|}{V(x)}.$$

4. Let (E,ρ) be a Polish space. Prove that under the total variation distance, the space $\mathscr{P}(E)$ is complete. Exemplify it may not be separable.

5. Let $(E,\mathscr{E})$ be a Polish space, and let $\mu,\nu\in\mathscr{P}(E)$. Construct a probability space $(\Omega,\mathscr{A},\mathbb{P})$ and measurable

$$\xi,\eta:\Omega\to E,$$

which are called random variables on E such that $\mathbb{P}\circ\xi^{-1}=\mu$, $\mathbb{P}\circ\eta^{-1}=\nu$ and $W_p(\mu,\nu)^p = \mathbb{E}\rho(\xi,\eta)^p$.

6. Let (E,ρ) be a Polish space and $\{\mu_n,\mu\}_{n\geqslant 1}\subset\mathscr{P}_p(E)$. Prove that $W_p(\mu_n,\mu)\to 0$ if and only if $\mu_n\xrightarrow{w}\mu$ and $\lim_{n\to\infty}\mu_n(\rho(o,\cdot)^p)=\mu_n(\rho(o,\cdot)^p)$, where $o\in E$ is a fixed point.

7. Let (E,ρ) be a compact metric space. Prove that for any $p\in[1,\infty)$, $(\mathscr{P}(E),W_p)$ is also a compact metric space. Exemplify that (E,ρ) is a locally compact space, but $(\mathscr{P}(E),W_p)$ is not.

8. (*Lévy distance*) For any probability distribution functions F,G on $\mathbb{R}$, let

$$\rho_L(F,G)=\inf\{\varepsilon\geqslant 0: F(x-\varepsilon)-\varepsilon\leqslant G(x)\leqslant F(x+\varepsilon)+\varepsilon,\ x\in\mathbb{R}^n\}.$$

Prove that ρ is a distance and $\rho(F_n,F)\to 0$ if and only if $F_n(x)\to F(x)$ for any continuous point x of F.

9. For any 1-dimensional random variables ξ,η, let F_ξ and F_η be their distribution functions. Define

$$\alpha(\xi,\eta)=\inf\{\varepsilon\geqslant 0:\mathbb{P}(|\xi-\eta|>\varepsilon)\leqslant\varepsilon\}$$

and

$$\beta(\xi,\eta)=\mathbb{E}\left(\frac{|\xi-\eta|}{1+|\xi-\eta|}\right).$$

Prove

$$\rho_L(F_\xi,F_\eta)\leqslant\alpha(\xi,\eta)$$

and

$$\frac{\alpha(\xi,\eta)^2}{1+\alpha(\xi,\eta)}\leqslant\beta(\xi,\eta)\leqslant\alpha(\xi,\eta)+\frac{(1-\alpha(\xi,\eta))\alpha(\xi,\eta)}{1+\alpha(\xi,\eta)}.$$

Chapter 8

Calculus on the Space of Finite Measures

In this chapter, we introduce the intrinsic and extrinsic derivatives for functions of finite measures and make corresponding calculus. For simplicity, we only consider measures on $\mathbb{R}^d$, but the related theory also applies to measures on more general spaces, such as Riemannian manifolds and separable Banach spaces.

Recall that $\mathfrak{M}$ is the set of all finite measures on $\mathbb{R}^d$. For fixed $k \in [1, \infty)$, let

$$\mathfrak{M}_k = \{\mu \in \mathfrak{M} : \ \mu(|\cdot|^k) < \infty\}$$

be the set of all finite measures on $\mathbb{R}^d$ having finite kth moment, and let

$$\mathscr{P}_k = \{\mu \in \mathfrak{M} : \ \mu(\mathbb{R}^d) = 1\}$$

be the set of probability measures on $\mathbb{R}^d$ having finite kth moment. Both are Polish space under the k-Wasserstein distance $\mathbb{W}_k$, which is defined in Definition 7.4 on $\mathscr{P}_k$, and for $\mu, \nu \in \mathfrak{M}$

$$\mathbb{W}_k(\mu, \nu) := \begin{cases} \|\mu\|_k := [\mu(|\cdot|^k)]^{\frac{1}{k}} & \text{if } \nu = 0, \\ \|\nu\|_k := [\mu(|\cdot|^k)]^{\frac{1}{k}} & \text{if } \mu = 0, \\ \mathbb{W}_k\left(\frac{\mu}{\mu(\mathbb{R}^d)}, \frac{\nu}{\nu(\mathbb{R}^d)}\right) + |\mu(\mathbb{R}^d) - \nu(\mathbb{R}^d)| & \text{otherwise.} \end{cases}$$

We will define intrinsic and extrinsic derivatives on $\mathscr{P}_k$ and $\mathfrak{M}_k$, and make calculus with these derivatives.

8.1 Intrinsic Derivative and Chain Rule

The intrinsic derivative for measures was introduced in [1] to construct diffusion processes on configuration spaces over a Riemannian manifold, and was used in [16] to study the geometry of dissipative evolution equations, see [2] for analysis and geometry on the Wasserstein space over a metric measure space.

In this chapter, we introduce the intrinsic derivative on $\mathscr{P}_k$ for $k \in [1, \infty)$ and establish the chain rule for functions of the distributions of random variables having kth moment.

8.1.1 Vector field and tangent space

To define the intrinsic derivative, let us first recall the directional derivative along a vector $v \in \mathbb{R}^d$ of a differentiable function f on $\mathbb{R}^d$:

$$\nabla_v f(x) := \lim_{\varepsilon \downarrow 0} \frac{f(x + \varepsilon v) - f(x)}{\varepsilon}, \quad x \in \mathbb{R}^d.$$

The directional derivative operator ∇_v reflects the variance rate of a function along the line

$$[0, 1] \ni \varepsilon \mapsto x + \varepsilon v,$$

which pushes forward a particle from point x along the direction v. Moreover, the gradient

$$\nabla f(x) := (\partial_{x_1} f(x), \ldots, \partial_{x_d} f(x))$$

is the element in $\mathbb{R}^d$ such that

$$\langle \nabla f(x), v \rangle = \nabla_v f(x), \quad v \in \mathbb{R}^d. \tag{8.1.1}$$

Now, let us characterize the variance rate of a function f on $\mathscr{P}_k$. In this case, we replace $x \in \mathbb{R}^d$ by a distribution $\mu \in \mathscr{P}_k$. To push forward the distribution μ, we need to push forward all points x in support of μ, so that instead of the line $\varepsilon \mapsto x + \varepsilon v$, we need a family of lines $\{\varepsilon \mapsto x + \varepsilon v(x)\}_{x \in \mathbb{R}^d}$, which leads to the notion of vector field.

Definition 8.1. A vector field on $\mathbb{R}^d$ is a measurable map

$$v : \mathbb{R}^d \ni x \mapsto v(x) \in \mathbb{R}^d.$$

Now, for each vector field v on $\mathbb{R}^d$, we may push forward a measure $\mu \in \mathfrak{M}$ along v as

$$[0,1] \ni \varepsilon \mapsto \mu \circ (\mathrm{id} + \varepsilon v)^{-1},$$

where $\mathrm{id} : x \mapsto x$ is the identity map, and for each $\varepsilon \in [0,1]$, $\mu \circ (\mathrm{id} + \varepsilon v)^{-1}$ is a finite measure on $\mathbb{R}^d$ defined as

$$(\mu \circ (\mathrm{id} + \varepsilon v)^{-1})(A) := \mu\big((\mathrm{id} + \varepsilon v)^{-1}(A)\big)$$

for $A \in \mathscr{B}(\mathbb{R}^d)$, the Borel σ-algebra on $\mathbb{R}^d$.

Then we may define the directional derivative long v for a function f of measures as follows:

$$\lim_{\varepsilon \downarrow 0} \frac{f(\mu \circ (\mathrm{id} + \varepsilon v)^{-1}) - f(\mu)}{\varepsilon},$$

provided the limit exists. When a function on $\mathfrak{M}_k$ is concerned, we need to assume that $\mu \circ (\mathrm{id} + \varepsilon v)^{-1} \in \mathfrak{M}_k$ for $\mu \in \mathfrak{M}_k$. It is easy to see that this is equivalent to

$$\int_{\mathbb{R}^d} |v|^k \, \mathrm{d}\mu < \infty,$$

since by the integral transformation (Theorem 3.27),

$$\int_{\mathbb{R}^d} |x|^k (\mu \circ (\mathrm{id} + \varepsilon v)^{-1})(\mathrm{d}x) = \int_{\mathbb{R}^d} |v|^k \, \mathrm{d}\mu.$$

This leads to the notion of tangent space at a point $\mu \in \mathfrak{M}_k$, which is the class of all vector fields on $\mathbb{R}^d$ such that $\mu(|v|^k) < \infty$.

Definition 8.2. Let $k \in [1, \infty)$ and $\mu \in \mathfrak{M}_k$. The tangent space at point μ is defined as

$$L^k(\mathbb{R}^d \to \mathbb{R}^d, \mu) := \{v : \mathbb{R}^d \to \mathbb{R}^d \text{ is measurable such that } \mu(|v|^k) < \infty\}.$$

We denote this space by $T_{\mu,k}$.

The tangent space $T_{\mu,k}$ is a separable Banach space, and when $k = 2$, it is a separable Hilbert space.

8.1.2 Intrinsic derivative and C^1 functions

We are now be able to define the directional derivative of a function on $\mathscr{P}_k$ (or $\mathfrak{M}_k$) along vector fields in the tangent space. Moreover, similar to (8.1.1), we may define the intrinsic derivative as a linear functional on the tangent space.

Definition 8.3. Let f be a continuous function on $\mathscr{P}_k$ (or $\mathfrak{M}_k$).

(1) Let $\mu \in \mathscr{P}_k$ (or $\mathfrak{M}_k$), and let $v \in T_{\mu,k}$. If the limit

$$D_v f(\mu) := \lim_{\varepsilon \downarrow 0} \frac{f(\mu \circ (\mathrm{id} + \varepsilon v)^{-1}) - f(\mu)}{\varepsilon}$$

exists, then it is called the directional derivative of f at μ along v.

(2) Let $\mu \in \mathscr{P}_k$ (or $\mathfrak{M}_k$). If $D_v f(\mu)$ exists for any $v \in T_{\mu,k}$, and

$$T_{\mu,k} \ni v \mapsto D_v f(\mu)$$

is a bounded linear functional, we call f intrinsically differentiable at μ. In this case, the linear functional

$$Df(\mu): \ v \mapsto D_v f(\mu)$$

is called the intrinsic derivative of f at μ.

(3) If f is intrinsically differentiable at all elements in $\mathscr{P}_k$ (or $\mathfrak{M}_k$), we call f intrinsically differentiable on $\mathscr{P}_k$ (or $\mathfrak{M}_k$).

According to the definition, for any intrinsically differentiable function f and any $\mu \in P_k$ (or $\mathfrak{M}_k$), $Df(\mu)$ is the unique element in

$$T_{\mu,k^*} := L^{k^*}(\mathbb{R}^d \to \mathbb{R}^d, \mu),$$

where $k^* := \frac{k}{k-1} \in (1, \infty]$, such that

$$D_v f(\mu) = \int_{\mathbb{R}^d} \langle Df(\mu)(x), v(x) \rangle \mu(\,\mathrm{d}x), \quad v \in T_{\mu,k}.$$

Intuitively, the intrinsic derivative describes the movement of distributions along the flows of particles induced by vector fields. As a random particle can be regarded as a random variable on $\mathbb{R}^d$, in the following, we lift a function f on $\mathscr{P}_k$ to a function $\hat{f}$ of random variables.

Let $\mathscr{R}_k$ be the class of all d-dimensional random variables ξ with $\mathbb{E}|\xi|^k < \infty$. Then a function f on $\mathscr{P}_k$ induces the following function on $\mathscr{R}_k$:

$$\mathscr{R}_k \ni \xi \mapsto \hat{f}(\xi) := f(\mathscr{L}_\xi),$$

where $\mathscr{L}_\xi \in \mathscr{P}_k$ is the distribution of ξ. The directional derivative of $\hat{f}$ at ξ along $\eta \in \mathscr{R}_k$ is defined as

$$\nabla_\eta \hat{f}(\xi) := \lim_{\varepsilon \downarrow 0} \frac{\hat{f}(\xi + \varepsilon\eta) - \hat{f}(\xi)}{\varepsilon},$$

provided the limit exists. We aim to establish the chain rule

$$\nabla_\eta \hat{f}(\xi) = \mathbb{E}[\langle Df(\mathscr{L}_\xi)(\xi), \eta \rangle], \quad \xi, \eta \in \mathscr{R}_k \tag{8.1.2}$$

for a class of intrinsically differentiable functions f on $\mathscr{P}_k$. To this end, we introduce the notion of L-derivative and the classes $C^1(\mathscr{P}_k)$ and $C_b^1(\mathscr{P}_k)$ as follows.

Definition 8.4. Let $f : \mathscr{P}_k \to \mathbb{R}$ be intrinsically differentiable.

(1) If

$$\lim_{\|\phi\|_{T_{\mu,k}} \downarrow 0} \frac{|f(\mu \circ (\mathrm{id} + \phi)^{-1}) - f(\mu) - D_\phi^I f(\mu)|}{\|\phi\|_{T_{\mu,k}}} = 0,$$

then f is called L-differentiable at μ. In this case, the intrinsic derivative is also called the L-derivative.

(2) We write $f \in C^1(\mathscr{P}_k)$ if f is L-differentiable at any $\mu \in \mathscr{P}_k$, and the L-derivative has a version $Df(\mu)(x)$ jointly continuous in $(x, \mu) \in \mathbb{R}^d \times \mathscr{P}_k$. If moreover $Df(\mu)(x)$ is bounded, we denote $f \in C_b^1(\mathscr{P}_k)$.

To establish the chain rule (8.1.2), we need to assume that the underlying probability space is Polish.

Definition 8.5. A probability space $(\Omega, \mathscr{F}, \mathbb{P})$ is called Polish if $\mathscr{F}$ is the $\mathbb{P}$-completeness of the a Borel σ-field induced by a Polish metric on Ω. $\mathbb{P}$ is called atomless if $\mathbb{P}(A) = 0$ holds for any atom $A \in \mathscr{F}$.

When $k = 2$, the L-derivative is named after Lions due to his Lecture Notes (see the corresponding reference in [6]), where $Df(\mu)$ is defined as the unique element in $T_{\mu,2}$ such that for any atomless probability space $(\Omega, \mathscr{F}, \mathbb{P})$ and any random variables X, Y with $\mathscr{L}_X = \mu$,

$$\lim_{\|Y - X\|_{L^2(\mathbb{P})} \downarrow 0} \frac{|f(\mathscr{L}_Y) - f(\mathscr{L}_X) - \mathbb{E}[\langle Df(\mu)(X), Y - X \rangle]|}{\|Y - X\|_{L^2(\mathbb{P})}} = 0.$$

Since $Df(\mu)$ does not depend on the choice of probability space, when μ is atomless, we may choose $(\Omega, \mathscr{F}, \mathbb{P}) = (\mathbb{R}^d, \mathscr{B}^d, \mu)$ such that this definition is equivalent to the one we introduced above. Since by approximations one may drop the atomless condition, so that the above definition of L-derivative coincides with, and is more straightforward than, the one defined by Lions.

8.1.3 Chain rule

To establish the chain rule (8.1.2) for functions of distributions of random variables, we need the following proposition.

Proposition 8.6. *Let $\{(\Omega_i, \mathscr{F}_i, \mathbb{P}_i)\}_{i=1,2}$ be two atomless, Polish probability spaces, and let $X_i, i = 1, 2$, be $\mathbb{R}^d$-valued random variables on these two probability spaces, respectively, such that $\mathscr{L}_{X_1|\mathbb{P}_1} = \mathscr{L}_{X_2|\mathbb{P}_2}$. Then for any $\varepsilon > 0$, there exist measurable maps*

$$\tau : \Omega_1 \to \Omega_2, \quad \tau^{-1} : \Omega_2 \to \Omega_1$$

such that

$$\begin{aligned}
&\mathbb{P}_1(\tau^{-1} \circ \tau = \mathrm{id}_{\Omega_1}) = \mathbb{P}_2(\tau \circ \tau^{-1} = \mathrm{id}_{\Omega_2}) = 1,\\
&\mathbb{P}_1 = \mathbb{P}_2 \circ \tau, \quad \mathbb{P}_2 = \mathbb{P}_1 \circ \tau^{-1},\\
&\|X_1 - X_2 \circ \tau\|_{L^\infty(\mathbb{P}_1)} + \|X_2 - X_1 \circ \tau^{-1}\|_{L^\infty(\mathbb{P}_2)} \leqslant \varepsilon,
\end{aligned}$$

where id_{Ω_i} stands for the identity map on $\Omega_i, i = 1, 2$.

Proof. Since $\mathbb{R}^d$ is separable, there is a measurable partition $(A_n)_{n\geqslant 1}$ of $\mathbb{R}^d$ such that $\mathrm{diam}(A_n) < \varepsilon$, $n \geqslant 1$. Let $A_n^i = \{X_i \in A_n\}, n \geqslant 1, i = 1, 2$. Then $(A_n^i)_{n\geqslant 1}$ forms a measurable partition of Ω_i so that $\sum_{n\geqslant 1} A_n^i = \Omega_i, i = 1, 2$, and, due to $\mathscr{L}_{X_1}|\mathbb{P}_1 = \mathscr{L}_{X_2}|\mathbb{P}_2$,

$$\mathbb{P}_1(A_n^1) = \mathbb{P}_2(A_n^2), \quad n \geqslant 1.$$

Since the probabilities $(\mathbb{P}_i)_{i=1,2}$ are atomless, according to Theorem C in Section 41 of [12], for any $n \geqslant 1$, there exist measurable sets $\tilde{A}_n^i \subset A_n^i$ with $\mathbb{P}_i(A_n^i \setminus \tilde{A}_n^i) = 0, i = 1, 2$, and a measurable bijective map

$$\tau_n : \tilde{A}_n^1 \to \tilde{A}_n^2$$

such that

$$\mathbb{P}_1|_{\tilde{A}_n^1} = \mathbb{P}_2 \circ \tau_n|_{\tilde{A}_n^1}, \quad \mathbb{P}_2|_{\tilde{A}_n^2} = \mathbb{P}_1 \circ \tau_n^{-1}|_{\tilde{A}_n^2}.$$

By $\text{diam}(A_n) < \varepsilon$ and $\mathbb{P}_i(A_n^i \setminus \tilde{A}_n^i) = 0$, we have

$$\|(X_1 - X_2 \circ \tau_n)1_{\tilde{A}_n^1}\|_{L^\infty(\mathbb{P}_1)} \vee \|(X_2 - X_1 \circ \tau_n^{-1})1_{\tilde{A}_n^2}\|_{L^\infty(\mathbb{P}_2)} \leqslant \varepsilon.$$

Then the proof is finished by taking, for fixed points $\hat{\omega}_i \in \Omega_i, i = 1, 2,$

$$\tau(\omega_1) = \begin{cases} \tau_n(\omega_1) & \text{if } \omega_1 \in \tilde{A}_n^1 \text{ for some } n \geqslant 1, \\ \hat{\omega}_2 & \text{otherwise,} \end{cases}$$

$$\tau^{-1}(\omega_2) = \begin{cases} \tau_n^{-1}(\omega_2) & \text{if } \omega_2 \in \tilde{A}_n^2 \text{ for some } n \geqslant 1, \\ \hat{\omega}_1 & \text{otherwise.} \end{cases}$$

□

The following chain rule is taken from Theorem 2.1 in [3], which extends the corresponding formulas for functions on $\mathscr{P}_2$ presented in [6, 13] and references within.

Theorem 8.7. *Let $f : \mathscr{P}_k \to \mathbb{R}$ be continuous for some $k \in [1, \infty)$, and let $(\xi_\varepsilon)_{\varepsilon \in [0,1]}$ be a family of $\mathbb{R}^d$-valued random variables on a complete probability space $(\Omega, \mathscr{F}, \mathbb{P})$ such that $\dot{\xi}_0 := \lim_{\varepsilon \downarrow 0} \frac{\xi_\varepsilon - \xi_0}{\varepsilon}$ exists in $L^k(\Omega \to \mathbb{R}^d, \mathbb{P})$. We assume that either ξ_ε is continuous in $\varepsilon \in [0, 1]$, or the probability space is Polish.*

(1) *Let $\mu_0 = \mathscr{L}_{\xi_0}$ be atomless. If f is L-differentiable such that $Df(\mu_0)$ has a continuous version satisfying*

$$\|Df(\mu_0)(x)\|_{\mathbb{R}^d} \leqslant C(1 + |x|^{k-1}), \quad x \in \mathbb{R}^d \tag{8.1.3}$$

for some constant $C > 0$, then

$$\lim_{\varepsilon \downarrow 0} \frac{f(\mathscr{L}_{\xi_\varepsilon}) - f(\mathscr{L}_{\xi_0})}{\varepsilon} = \mathbb{E}[\langle Df(\mu_0)(\xi_0), \dot{\xi}_0 \rangle]. \tag{8.1.4}$$

(2) *If f is L-differentiable in a neighborhood O of μ_0 such that Df has a version jointly continuous in $(x, \mu) \in \mathbb{R}^d \times O$ satisfying*

$$\|Df(\mu)(x)\|_{\mathbb{R}^d} \leqslant C(1 + |x|^{k-1}), \quad (x, \mu) \in \mathbb{R}^d \times O \tag{8.1.5}$$

for some constant $C > 0$, then (8.1.4) holds.

Proof. Without loss of generality, we may and do assume that $\mathbb{P}$ is atomless. Otherwise, by taking

$$(\tilde{\Omega}, \tilde{\mathscr{F}}, \tilde{\mathbb{P}}) = (\Omega \times [0,1], \mathscr{F} \times \mathscr{B}([0,1]), \mathbb{P} \times \mathrm{d}s),$$
$$(\tilde{\xi_\varepsilon})(\omega, s) = \xi_\varepsilon(\omega) \text{ for } (\omega, s) \in \tilde{\Omega},$$

where $\mathscr{B}([0,1])$ is the completion of the Borel σ-algebra on $[0,1]$, with respect to the Lebesgue measure $\mathrm{d}s$, we have

$$\mathscr{L}_{\tilde{\xi}_\varepsilon|\tilde{\mathbb{P}}} = \mathscr{L}_{\xi_\varepsilon|\mathbb{P}}, \quad \mathbb{E}[\langle Df(\mu_0)(\xi_0), \dot{\xi}_0\rangle] = \tilde{\mathbb{E}}[\langle Df(\mu_0)(\tilde{\xi}_0), \dot{\tilde{\xi}}_0\rangle].$$

In this way, we go back to the atomless situation. Moreover, it suffices to prove for the Polish probability space case. Indeed, when ξ_ε is continuous in ε, we may take $\bar{\Omega} = C([0,1];\mathbb{R}^d)$, let $\bar{\mathbb{P}}$ be the distribution of $\xi_\cdot$, let $\bar{\mathscr{F}}$ be the $\bar{\mathbb{P}}$-complete Borel σ-field on $\bar{\Omega}$ induced by the uniform norm, and consider the coordinate random variable $\bar{\xi}_\cdot(\omega) := \omega, \omega \in \bar{\Omega}$. Then

$$\mathscr{L}_{\bar{\xi}_\cdot|\tilde{\mathbb{P}}} = \mathscr{L}_{\xi_\cdot|\mathbb{P}},$$

so that $\mathscr{L}_{\bar{\xi}_\varepsilon|\bar{\mathbb{P}}} = \mathscr{L}_{\xi_\varepsilon|\mathbb{P}}$ for any $\varepsilon \in [0,1]$ and $\mathscr{L}_{\bar{\xi}_0'|\bar{\mathbb{P}}} = \mathscr{L}_{\xi_0'|\mathbb{P}}$, hence we have reduced the situation to the Polish setting.

(1) Let $\mathscr{L}_{\xi_0} = \mu_0 \in \mathscr{P}_k$ be atomless. In this case, $(\mathbb{R}^d, \mathscr{B}(\mathbb{R}^d), \mu_0)$ is an atomless Polish complete probability space, where $\mathscr{B}(\mathbb{R}^d)$ is the μ_0-complete Borel σ-algebra of $\mathbb{R}^d$. By Proposition 8.6, for any $n \geqslant 1$, we find measurable maps

$$\tau_n : \Omega \to \mathbb{R}^d, \quad \tau_n^{-1} : \mathbb{R}^d \to \Omega$$

such that

$$\begin{aligned} &\mathbb{P}(\tau_n^{-1} \circ \tau_n = \mathrm{id}_\Omega) = \mu_0(\tau_n \circ \tau_n^{-1} = \mathrm{id}) = 1, \\ &\mathbb{P} = \mu_0 \circ \tau_n, \quad \mu_0 = \mathbb{P} \circ \tau_n^{-1}, \\ &\|\xi_0 - \tau_n\|_{L^\infty(\mathbb{P})} + \|\mathrm{id} - \xi_0 \circ \tau_n^{-1}\|_{L^\infty(\mu_0)} \leqslant \frac{1}{n}, \end{aligned} \tag{8.1.6}$$

where id_Ω is the identity map on Ω.

Since f is L-differentiable at μ_0, there exists a decreasing function $h : [0,1] \to [0,\infty)$ with $h(r) \downarrow 0$ as $r \downarrow 0$ such that

$$\begin{aligned} &\sup_{\|\phi\|_{L^k(\mu_0)} \leqslant r} \left| f(\mu_0 \circ (\mathrm{id} + \phi)^{-1}) - f(\mu_0) - D_\phi f(\mu_0) \right| \\ &\leqslant rh(r), \quad r \in [0,1]. \end{aligned} \tag{8.1.7}$$

By $\mathscr{L}_{\xi_\varepsilon-\xi_0} \in \mathscr{P}_k$ and (8.1.6), we have

$$\phi_{n,\varepsilon} := (\xi_\varepsilon - \xi_0) \circ \tau_n^{-1} \in T_{\mu,k}, \quad \|\phi_{n,\varepsilon}\|_{T_{\mu,k}} = \|\xi_\varepsilon - \xi_0\|_{L^k(\mathbb{P})}. \quad (8.1.8)$$

Next, (8.1.6) implies

$$\begin{aligned}\mathscr{L}_{\tau_n+\xi_\varepsilon-\xi_0} &= \mathbb{P} \circ (\tau_n + \xi_\varepsilon - \xi_0)^{-1} \\ &= (\mu_0 \circ \tau_n) \circ (\tau_n + \xi_\varepsilon - \xi_0)^{-1} = \mu_0 \circ (\mathrm{id} + \phi_{n,\varepsilon})^{-1}. \quad (8.1.9)\end{aligned}$$

Moreover, by $\frac{\xi_\varepsilon-\xi_0}{\varepsilon} \to \dot{\xi}_0$ in $L^k(\mathbb{P})$ as $\varepsilon \downarrow 0$, we find a constant $c \geqslant 1$ such that

$$\|\xi_\varepsilon - \xi_0\|_{L^k(\mathbb{P})} \leqslant c\varepsilon, \quad \varepsilon \in [0,1]. \quad (8.1.10)$$

Combining (8.1.6)–(8.1.10) leads to

$$\begin{aligned}&|f(\mathscr{L}_{\tau_n+\xi_\varepsilon-\xi_0}) - f(\mathscr{L}_{\xi_0}) - \mathbb{E}[\langle (Df)(\mu_0)(\tau_n), (\xi_\varepsilon - \xi_0)\rangle]| \\ &= \left|f(\mu_0 \circ (\mathrm{id} + \phi_{n,\varepsilon})^{-1}) - f(\mu_0) - D_{\phi_{n,\varepsilon}} f(\mu_0)\right| \\ &\leqslant \|\phi_{n,\varepsilon}\|_{T_{\mu,k}} h(\|\phi_{n,\varepsilon}\|_{T_{\mu,k}}) \\ &= \|\xi_\varepsilon - \xi_0\|_{L^k(\mathbb{P})} h(\|\xi_\varepsilon - \xi_0\|_{L^k(\mathbb{P})}), \quad \varepsilon \in [0, c^{-1}]. \quad (8.1.11)\end{aligned}$$

Since $f(\mu)$ is continuous in μ and $Df(\mu_0)(x)$ is continuous in x, by (8.1.3) and (8.1.6), we may apply the dominated convergence theorem to deduce from (8.1.11) with $n \to \infty$ that

$$\begin{aligned}&\left|f(\mathscr{L}_{\xi_\varepsilon}) - f(\mathscr{L}_{\xi_0}) - \mathbb{E}[\langle (Df)(\mu_0)(\xi_0), (\xi_\varepsilon - \xi_0)\rangle]\right| \\ &\leqslant \|\xi_\varepsilon - \xi_0\|_{L^k(\mathbb{P})} h(\|\xi_\varepsilon - \xi_0\|_{L^k(\mathbb{P})}), \quad \varepsilon \in [0, c^{-1}].\end{aligned}$$

Combining this with (8.1.10) and $h(r) \to 0$ as $r \to 0$, we derive (8.1.4).

(2) When μ_0 has an atom, we take an $\mathbb{R}^d$-valued bounded random variable X which is independent of $(\xi_\varepsilon)_{\varepsilon\in[0,1]}$ and $\mathscr{L}_X$ does not have any atom. Then

$$\mathscr{L}_{\xi_0+sX+r(\xi_\varepsilon-\xi_0)} \in \mathscr{P}_k$$

does not have atom for any $s > 0, \varepsilon \in [0,1]$. By conditions in Theorem 8.7(2), there exists a small constant $s_0 \in (0,1)$ such that for any

$s, \varepsilon \in (0, s_0]$, we may apply (8.1.4) to the random variables

$$\xi_0 + sX + (r+\delta)(\xi_\varepsilon - \xi_0), \quad \delta > 0$$

to conclude

$$\begin{aligned} & f(\mathscr{L}_{\xi_\varepsilon + sX}) - f(\mathscr{L}_{\xi_0 + sX}) \\ &= \int_0^1 \frac{\mathrm{d}}{\mathrm{d}\delta} f(\mathscr{L}_{\xi_0 + sX + (r+\delta)(\xi_\varepsilon - \xi_0)})\big|_{\delta=0} \,\mathrm{d}r \\ &= \int_0^1 \mathbb{E}[\langle Df(\mathscr{L}_{\xi_0 + sX + r(\xi_\varepsilon - \xi_0)})(\xi_0 + sX + r(\xi_\varepsilon - \xi_0)), \xi_\varepsilon - \xi_0\rangle] \,\mathrm{d}r. \end{aligned}$$

By conditions in Theorem 8.7(2), we may let $s \downarrow 0$ to derive

$$\begin{aligned} & f(\mathscr{L}_{\xi_\varepsilon}) - f(\mathscr{L}_{\xi_0}) \\ &= \int_0^1 \mathbb{E}[\langle Df(\mathscr{L}_{\xi_0 + r(\xi_\varepsilon - \xi_0)})(\xi_0 + r(\xi_\varepsilon - \xi_0)), \xi_\varepsilon - \xi_0\rangle] \,\mathrm{d}r, \\ & \varepsilon \in (0, s_0). \end{aligned}$$

Multiplying both sides by ε^{-1} and letting $\varepsilon \downarrow 0$, we finish the proof. □

As a consequence of the chain rule, we have the following Lipschitz estimate for L-differentiable functions on $\mathscr{P}_k$.

Corollary 8.8. *Let f be L-differentiable on $\mathscr{P}_k$ such that for any $\mu \in \mathscr{P}_k$, $Df(\mu)(\cdot)$ has a continuous version satisfying*

$$|Df(\mu)(x)| \leqslant c(\mu)(1 + |x|^{k-1}), \quad x \in \mathbb{R}^d \tag{8.1.12}$$

for come constant $c(\mu) > 0$, and

$$K_0 := \sup_{\mu \in \mathscr{P}_k} \|Df(\mu)\|_{L^k(\mu)} < \infty. \tag{8.1.13}$$

Then

$$|f(\mu_1) - f(\mu_2)| \leqslant K_0 \mathbb{W}_k(\mu_1, \mu_2), \quad \mu_1, \mu_2 \in \mathscr{P}_k. \tag{8.1.14}$$

Proof. By Theorem 7.4, there exists $\pi \in \mathscr{C}(\mu_1, \mu_2)$ such that

$$\mathbb{W}_k(\mu_1, \mu_2) = \left(\int_{\mathbb{R}^d \times \mathbb{R}^d} |x - y|^k \pi(\mathrm{d}x, \mathrm{d}y) \right)^{\frac{1}{k}}.$$

Now, consider the probability space

$$(\Omega, \mathscr{F}, \mathbb{P}) = (\mathbb{R}^d \times \mathbb{R}^d \times \mathbb{R}^d, \bar{\mathscr{B}}(\mathbb{R}^d \times \mathbb{R}^d \times \mathbb{R}^d), \pi \times G),$$

where G is the standard Gaussian measure on $\mathbb{R}^d$ and $\bar{\mathscr{B}}(\mathbb{R}^d \times \mathbb{R}^d \times \mathbb{R}^d)$ is the completion of the Borel σ-field $\mathscr{B}(\mathbb{R}^d \times \mathbb{R}^d \times \mathbb{R}^d)$ with respect to $\mathbb{P}$. Obviously, this probability space is atomless and Polish, and the random variables

$$\xi_1(\omega) := \omega_1, \quad \xi_2(\omega) := \omega_2, \quad \omega = (\omega_1, \omega_2, \omega_3) \in \Omega := \mathbb{R}^d \times \mathbb{R}^d \times \mathbb{R}^d$$

satisfy

$$\mathscr{L}_{\xi_1} = \mu_1, \quad \mathscr{L}_{\xi_2} = \mu_2, \quad \mathbb{W}_k(\mu_1, \mu_2) = (\mathbb{E}[|\xi_1 - \xi_2|^k])^{\frac{1}{k}}.$$

Moreover, the random variable

$$\eta(\omega) := \omega_3, \quad \omega = (\omega_1, \omega_2, \omega_3) \in \Omega$$

is independent of (ξ_1, ξ_2) with distribution $\mathscr{L}_\mu = G$, so that the random variables

$$\gamma_\varepsilon(r) := \varepsilon\eta + r\xi_1 + (1-r)\xi_2, \quad r \in [0,1], \varepsilon \in (0,1]$$

are absolutely continuous with respect to the Lebesgue measure. By Theorem 8.7, (8.1.12) and the continuity of $Df(\mu)(\cdot)$, we obtain

$$\begin{aligned}
|f(\mathscr{L}_{\gamma_\varepsilon(1)}) - f(\mathscr{L}_{\gamma_\varepsilon(0)})| &= \left| \int_0^1 \mathbb{E}\big[\langle Df(\mathscr{L}_{\gamma_\varepsilon(r)})(\gamma_\varepsilon(r)), \xi_1 - \xi_2\rangle\big]\, \mathrm{d}r \right| \\
&\leqslant (\mathbb{E}[|\xi_1 - \xi_2|^k])^{\frac{1}{k}} \int_0^1 \|Df(\mathscr{L}_{\gamma_\varepsilon(r)})\|_{L^k(\mathscr{L}_{\gamma_\varepsilon(r)})}\, \mathrm{d}r \\
&\leqslant K\mathbb{W}_k(\mu_1, \mu_2), \quad \varepsilon \in (0,1].
\end{aligned}$$

Letting $\varepsilon \to 0$, we derive (8.1.14). □

8.2 Extrinsic Derivative and Convexity Extrinsic Derivative

Regarding a measure as the distribution of particle systems, the intrinsic derivative describes the movement of particles. In this part, we consider the (convexity) extrinsic derivative, which refers to the birth and death rates of particles.

We first recall the extrinsic derivative defined as partial derivative in the direction of Dirac measures, see [17, Definition 1.2].

Definition 8.9 (Extrinsic derivative). Let f be a real function on $\mathfrak{M}_k$. For any $x \in \mathbb{R}^d$, let δ_x be the Dirac measure at x, i.e. $\delta_x \in \mathscr{P}$ with $\delta_x(\{x\}) = 1$.

(1) f is called extrinsically differentiable on $\mathfrak{M}_k$ with derivative $D^E f$ if

$$D^E f(\mu)(x) := \lim_{\varepsilon \downarrow 0} \frac{f(\mu + \varepsilon\delta_x) - f(\mu)}{\varepsilon} \in \mathbb{R}$$

exists for all $(x, \mu) \in \mathbb{R}^d \times \mathfrak{M}_k$.

(2) If $D^E f(\mu)(x)$ exists and is continuous in $(x, \mu) \in \mathbb{R}^d \times \mathfrak{M}_k$, we denote $f \in C^{E,1}(\mathfrak{M}_k)$.

(3) We denote $f \in C_K^{E,1}(\mathfrak{M}_k)$ if $f \in C^{E,1}(\mathfrak{M}_k)$ and for any compact set $\mathscr{K} \subset \mathfrak{M}_k$, there exists a constant $C > 0$ such that

$$\sup_{\mu \in \mathscr{K}} |D^E f(\mu)(x)| \leqslant C(1 + |x|^k), \quad x \in \mathbb{R}^d.$$

(4) We denote $f \in C^{E,1,1}(\mathfrak{M}_k)$ if $f \in C^{E,1}(\mathfrak{M}_k)$ such that $D^E f(\mu)(x)$ is differentiable in x, $\nabla\{D^E f(\mu)(\cdot)\}(x)$ is continuous in $(x, \mu) \in \mathbb{R}^d \times \mathfrak{M}_k$ and $|\nabla\{D^E f(\mu)\}| \in L^{\frac{k}{k-1}}(\mu)$ for any $\mu \in \mathfrak{M}_k$.

(5) We write $f \in C_B^{E,1,1}(\mathfrak{M}_k)$ if $f \in C^{E,1,1}(\mathfrak{M}_k)$ and for any constant $L > 0$, there exists $C_L > 0$ such that

$$\sup_{\|\mu\|_k \leqslant L} |\nabla\{D^E f(\mu)\}|(x) \leqslant C_L(1 + |x|^k), \quad x \in \mathbb{R}^d.$$

Since for a probability measure μ and $s > 0$, $\mu + s\delta_x$ is no longer a probability measure, for functions of probability measures, we modify the definition of the extrinsic derivative with the convex combination

$$(1 - s)\mu + s\delta_x,$$

replacing $\mu + s\delta_x$. This leads to the notion of convexity extrinsic derivative.

Definition 8.10 (Convexity extrinsic derivative). Let f be a real function on $\mathscr{P}_k$.

(1) f is called extrinsically differentiable on $\mathscr{P}_k$ if the centered extrinsic derivative

$$\tilde{D}^E f(\mu)(x) := \lim_{s \downarrow 0} \frac{f((1 - s)\mu + s\delta_x) - f(\mu)}{s} \in \mathbb{R}$$

exists for all $(x, \mu) \in \mathbb{R}^d \times \mathscr{P}_k$.

(2) We write $f \in C^{E,1}(\mathscr{P}_k)$, if $\tilde{D}^E f(\mu)(x)$ exists and is continuous in $(x, \mu) \in \mathbb{R}^d \times \mathscr{P}_k$.
(3) We denote $f \in C_K^{E,1}(\mathscr{P}_k)$, if $f \in C^{E,1}(\mathscr{P}_k)$ and for any compact set $\mathscr{K} \subset \mathscr{P}_k$, there exists a constant $C > 0$ such that

$$\sup_{\mu \in \mathscr{K}} |\tilde{D}^E f(\mu)(x)| \leqslant C(1 + |x|^k), \quad x \in \mathbb{R}^d.$$

(4) We write $f \in C^{E,1,1}(\mathscr{P}_k)$ if $f \in C^{E,1}(\mathscr{P}_k)$ such that $\tilde{D}^E f(\mu)(x)$ is differentiable in $x \in \mathbb{R}^d$, $\nabla\{\tilde{D}^E f(\mu)\}(x)$ is continuous in $(x, \mu) \in \mathbb{R}^d \times \mathscr{P}_k$, and $|\nabla\{\tilde{D}^E f(\mu)\}| \in L^{\frac{k}{k-1}}(\mu)$ for any $\mu \in \mathscr{P}_k$.
(5) We write $f \in C_B^{E,1,1}(\mathscr{P}_k)$, if $f \in C^{E,1,1}(\mathscr{P}_k)$, and for any constant $L > 0$, there exists $C > 0$ such that

$$\sup_{\mu(|\cdot|^k) \leqslant L} |\nabla\{\tilde{D}^E f(\mu)\}|(x) \leqslant C(1 + |x|^k), \quad x \in \mathbb{R}^d.$$

By Proposition 8.13 with $\gamma = \delta_x$ and $r = 0$, we have

$$\lim_{s \downarrow 0} \frac{f((1-s)\mu + s\delta_x) - f(\mu)}{s}$$
$$= D^E f(\mu)(x) - \mu(D^E f(\mu)), \quad f \in C_K^{E,1}(\mathfrak{M}_k), \ x \in \mathbb{R}^d.$$

So, the convexity extrinsic derivative is indeed the centralized extrinsic derivative.

For $\mu \in \mathfrak{M}$ and a density function $0 \leqslant h \in L^1(\mu)$, $h\mu$ is a finite measure on $\mathbb{R}^d$ defined as

$$(h\mu)(A) = \int_A h \, \mathrm{d}\mu, \quad A \in \mathscr{B}(\mathbb{R}^d).$$

Then a function f on $\mathfrak{M}$ induces the following function of density h:

$$h \mapsto f(h\mu).$$

To characterize this function by using the extrinsic derivative, we introduce the following class of density functions.

Definition 8.11. We denote $h \in \mathscr{H}_{\varepsilon_0}$ for a constant $\varepsilon_0 > 0$ if h satisfies the following conditions:

(1) $0 \leqslant h \in C([0, \varepsilon_0] \times \mathbb{R}^d)$;
(2) $h_0 \equiv 0$, $\sup_{\varepsilon \in [0,\varepsilon_0]} \|h_\varepsilon\|_\infty < \infty$, and there exists a compact set $K \subset \mathbb{R}^d$ such that $h_\varepsilon|_{K^c} = 0$ for all $\varepsilon \in [0, \varepsilon_0]$;

(3) $\dot{h}_\varepsilon := \lim_{s\downarrow 0} \frac{h_{\varepsilon+s}-h_\varepsilon}{s} \in C_b(\mathbb{R}^d)$ exists and is uniformly bounded for $\varepsilon \in [0, \varepsilon_0)$.

The following proposition links $f((1+h_\varepsilon)\mu)-f(\mu)$ to the extrinsic derivative, which will be used to characterize the relation of Df and $D^E f$ in the following section.

Proposition 8.12. *Let $k \in [1, \infty)$. For any $h \in \mathscr{H}_{\varepsilon_0}$ and any $f \in C^{E,1,1}(\mathfrak{M}_k)$,*

$$f((1+h_\varepsilon)\mu) - f(\mu) = \int_0^\varepsilon \mathrm{d}r \int_{\mathbb{R}^d} D^E f((1+h_r)\mu)(x)\dot{h}_r(x)\mu(\mathrm{d}x) \tag{8.2.1}$$

holds for all $\mu \in \mathfrak{M}_k$ and $\varepsilon \in [0, \varepsilon_0]$.

Proof. (1) We first consider

$$\mu \in \mathfrak{M}_{\text{disc}} := \left\{ \sum_{i=1}^n a_i \delta_{x_i} : n \geqslant 1, a_i > 0, x_i \in \mathbb{R}^d, 1 \leqslant i \leqslant n \right\}.$$

In this case, for any $\varepsilon \in [0, \varepsilon_0)$ and $s \in (0, \varepsilon_0 - \varepsilon)$, by the definition of D^E, we have

$$\begin{aligned}
&f((1+h_{\varepsilon+s})\mu) - f((1+h_\varepsilon)\mu) \\
&= f\left((1+h_\varepsilon)\mu + \sum_{i=1}^n \{h_{\varepsilon+s} - h_\varepsilon\}(x_i)a_i\delta_{x_i} \right) - f((1+h_\varepsilon)\mu) \\
&= \sum_{k=1}^n \Bigg\{ f\left((1+h_\varepsilon)\mu + \sum_{i=1}^k \{h_{\varepsilon+s} - h_\varepsilon\}(x_i)a_i\delta_{x_i} \right) \\
&\qquad - f\left((1+h_\varepsilon)\mu + \sum_{i=1}^{k-1} \{h_{\varepsilon+s} - h_\varepsilon\}(x_i)a_i\delta_{x_i} \right) \Bigg\} \\
&= \sum_{k=1}^n \int_0^{a_k\{h_{\varepsilon+s}-h_\varepsilon\}(x_k)} \\
&\quad \left\{ D^E f\left((1+h_\varepsilon)\mu + \sum_{i=1}^{k-1} \{h_{\varepsilon+s} - h_\varepsilon\}(x_i)a_i\delta_{x_i} + r\delta_{x_k} \right) \right\} (x_k)\,\mathrm{d}r.
\end{aligned}$$

Multiplying by s^{-1} and letting $s \downarrow 0$, we deduce from this and the continuity of $D^E f$ that

$$\begin{aligned}
&\lim_{s\downarrow 0} \frac{f((1+h_{\varepsilon+s})\mu) - f((1+h_\varepsilon)\mu)}{s} \\
&= \sum_{k=1}^{n} a_k \dot{h}_\varepsilon(x_k) D^E f((1+h_\varepsilon)\mu)(x_k) \\
&= \int_{\mathbb{R}^d} D^E f((1+h_\varepsilon)\mu)(x)\dot{h}_\varepsilon(x)\mu(\mathrm{d}x), \quad \varepsilon \in [0,\varepsilon_0), \mu \in \mathfrak{M}_{\mathrm{disc}}.
\end{aligned} \tag{8.2.2}$$

(2) In general, for any $\mu \in \mathfrak{M}_k$, let $\{\mu_n\}_{n\geqslant 1} \subset \mathfrak{M}_{\mathrm{disc}}$ such that $\mu_n \to \mu$ in $\mathfrak{M}_k$. By (8.2.2), for any $\varepsilon \in (0,\varepsilon_0)$ and $s \in (0, \varepsilon_0 - \varepsilon)$, we have

$$\begin{aligned}
&f((1+h_\varepsilon)\mu_n) - f(\mu_n) \\
&= \int_0^\varepsilon \mathrm{d}r \int_{\mathbb{R}^d} D^E f((1+h_r)\mu_n)(x)\dot{h}_r(x)\mu_n(\mathrm{d}x), \quad n \geqslant 1.
\end{aligned} \tag{8.2.3}$$

Next, since $D^E f \in C(\mathbb{R}^d \times \mathfrak{M}_k)$ and $h_r, \dot{h}_r \in C_b(\mathbb{R}^d)$ for $r \in [0,\varepsilon_0]$ with compact support $\subset K$, and $\mu_n \to \mu$ in $\mathfrak{M}_k$, we obtain

$$\begin{aligned}
&\lim_{n\to\infty} \int_{\mathbb{R}^d} D^E f((1+h_r)\mu)(x)\dot{h}_r(x)\mu_n(\mathrm{d}x) \\
&= \int_{\mathbb{R}^d} D^E f((1+h_r)\mu)(x)\dot{h}_r(x)\mu(\mathrm{d}x).
\end{aligned} \tag{8.2.4}$$

Moreover, $\mu_n \to \mu$ in $\mathfrak{M}_k$ and $h \in \mathscr{H}_{\varepsilon_0}$ imply that the set

$$\mathscr{K}_r := \{(1+h_r)\mu, (1+h_r)\mu_n : n \geqslant 1\}$$

is compact in $\mathfrak{M}_k$ for any $r \in [0,\varepsilon_0]$. Combining this with $D^E f \in C(\mathbb{R}^d \times \mathfrak{M}_k)$, we see that the function

$$\mathscr{K}_r \times \mathbb{R}^d \ni (\gamma, x) \mapsto D^E f(\gamma)(x)\dot{h}_r(x)$$

is uniformly continuous and has compact support $\subset \mathscr{K}_r \times K$, so that (8.2.4) implies

$$\begin{aligned}
&\limsup_{n\to\infty}\bigg|\int_{\mathbb{R}^d} D^E f((1+h_r)\mu_n)(x)\dot{h}_r(x)\mu_n(\mathrm{d}x) \\
&\qquad - \int_{\mathbb{R}^d} D^E f((1+h_r)\mu)(x)\dot{h}_r(x)\mu(\mathrm{d}x)\bigg| \\
&= \limsup_{n\to\infty}\bigg|\int_{\mathbb{R}^d} D^E f((1+h_r)\mu_n)(x)\dot{h}_r(x)\mu_n(\mathrm{d}x) \\
&\qquad - \int_{\mathbb{R}^d} D^E f((1+h_r)\mu)(x)\dot{h}_r(x)\mu_n(\mathrm{d}x)\bigg| \\
&\leqslant \limsup_{n\to\infty}\bigg\{\mu_n(K)\sup_{x\in K}|D^E f((1+h_r)\mu_n)(x)\dot{h}_r(x) \\
&\qquad - D^E f((1+h_r)\mu)(x)\dot{h}_r(x)|\bigg\} \\
&= 0.
\end{aligned}$$

Combining this with

$$\sup_{(\gamma,x)\in\mathscr{K}_r\times K, r\in[0,\varepsilon_0]} |D^E f(\gamma)(x)\dot{h}_r(x)| < \infty,$$

we deduce from the dominated convergence theorem that

$$\begin{aligned}
&\lim_{n\to\infty}\int_0^\varepsilon \mathrm{d}r\int_{\mathbb{R}^d}\{D^E f\}((1+h_\varepsilon)\mu_n)(x)\dot{h}_r(x)\mu_n(\mathrm{d}x) \\
&= \int_0^\varepsilon \mathrm{d}r\int_{\mathbb{R}^d}\{D^E f\}((1+h_r)\mu)(x)\dot{h}_r(x)\mu(\mathrm{d}x). \qquad (8.2.5)
\end{aligned}$$

Therefore, by letting $n\to\infty$ in (8.2.3) and using the continuity of f, we prove (8.2.1). □

To calculate the convexity extrinsic derivative, we present the following result.

Proposition 8.13. *Let $k \in [1, \infty)$. Then for any $f \in C_K^{E,1}(\mathfrak{M}_k)$ and $\mu, \gamma \in \mathfrak{M}_k$,*

$$\begin{aligned} &\frac{\mathrm{d}}{\mathrm{d}r} f((1-r)\mu + r\gamma) \\ &\quad := \lim_{\varepsilon \downarrow 0} \frac{f((1-r-\varepsilon)\mu + (r+\varepsilon)\gamma) - f((1-r)\mu + r\gamma)}{\varepsilon} \\ &\quad = \int_{\mathbb{R}^d} \{D^E f((1-r)\mu + r\gamma)(x)\}(\gamma - \mu)(\mathrm{d}x), \quad r \in [0,1). \end{aligned}$$

Consequently, for any $f \in C_K^{E,1}(\mathfrak{M}_k)$,

$$\begin{aligned} &\tilde{D}^E f(\mu)(x) \\ &\quad := \lim_{s \downarrow 0} \frac{f((1-s)\mu + s\delta_x) - f(\mu)}{s} \\ &\quad = D^E f(\mu)(x) - \mu\big(D^E f(\mu)\big), \quad (x, \mu) \in \mathbb{R}^d \times \mathfrak{M}_k. \end{aligned}$$

The assertions also hold for $\mathscr{P}_k$ replacing $\mathfrak{M}_k$.

Proof. As in the proof of Proposition 8.12, we take

$$\mu_n = \sum_{i=1}^n \alpha_{n,i}\delta_{x_{n,i}}, \quad \gamma_n = \sum_{i=1}^n \beta_{n,i}\delta_{x_{n,i}}$$

for some $x_{n,i} \in \mathbb{R}^d$ and $\alpha_{n,i}, \beta_{n,i} \geqslant 0$, such that

$$\mu_n \to \mu, \quad \gamma_n \to \gamma \quad \text{in } \mathfrak{M}_k \quad \text{as } n \to \infty.$$

For any $r \in [0,1)$ and $\varepsilon \in (0, 1-r)$, let

$$\Lambda_{n,i}^\varepsilon := (1-r)\mu_n + r\gamma_n + \sum_{k=1}^{i-1} \varepsilon(\beta_k - \alpha_k)\delta_{x_{n,k}} \in \mathfrak{M}_k, \quad 1 \leqslant i \leqslant n,$$

where by convention $\sum_{i=1}^0 := 0$. Then by the definition of $D^E f$, we have

$$\begin{aligned} &f((1-r-\varepsilon)\mu_n + (r+\varepsilon)\gamma_n) - f((1-r)\mu_n + r\gamma_n) \\ &\quad = \sum_{i=1}^n \big\{f(\Lambda_{n,i}^\varepsilon + \varepsilon(\beta_{n,i} - \alpha_{n,i})\delta_{x_{n,i}}) - f(\Lambda_{n,i}^\varepsilon)\big\} \\ &\quad = \sum_{i=1}^n \int_0^{\varepsilon(\beta_{n,i} - \alpha_{n,i})} D^E f(\Lambda_{n,i}^\varepsilon + s\delta_{x_{n,i}})(x_{n,i})\, \mathrm{d}s, \quad \varepsilon \in (0, 1-r). \end{aligned}$$

Multiplying by ε^{-1} and letting $\varepsilon \downarrow 0$, due to the continuity of $D^E f$, we derive

$$\begin{aligned}
&\frac{\mathrm{d}}{\mathrm{d}r} f((1-r)\mu_n + r\gamma_n) \\
&\quad = \sum_{i=1}^{n} (\beta_{n,i} - \alpha_{n,i}) D^E f((1-r)\mu_n + r\gamma_n)(x_{n,i}) \\
&\quad = \int_{\mathbb{R}^d} \{D^E f((1-r)\mu_n + r\gamma_n)(x)\}(\gamma_n - \mu_n)(\mathrm{d}x), \\
&\quad r \in [0,1), \quad n \geqslant 1.
\end{aligned}$$

Consequently, for any $r \in [0,1)$,

$$\begin{aligned}
&f((1-r-\varepsilon)\mu_n + (r+\varepsilon)\gamma_n) - f((1-r)\mu_n + r\gamma_n) \\
&\quad = \int_r^{r+\varepsilon} \mathrm{d}s \int_{\mathbb{R}^d} \{D^E f((1-s)\mu_n + s\gamma_n)(x)\}(\gamma_n - \mu_n)(\mathrm{d}x), \\
&\quad \varepsilon \in (0, 1-r), \quad n \geqslant 1.
\end{aligned}$$

Noting that the set $\{\mu_n, \gamma_n : n \geqslant 1\}$ is relatively compact in $\mathfrak{M}_k$, by this and the condition on f, we may let $n \to \infty$ to derive

$$\begin{aligned}
&f((1-r-\varepsilon)\mu + (r+\varepsilon)\gamma) - f((1-r)\mu + r\gamma) \\
&\quad = \int_r^{r+\varepsilon} \mathrm{d}s \int_{\mathbb{R}^d} \{D^E f((1-s)\mu + s\gamma)(x)\}(\gamma - \mu)(\mathrm{d}x), \\
&\quad \varepsilon \in (0, 1-r).
\end{aligned}$$

Multiplying by ε^{-1} and letting $\varepsilon \downarrow 0$, we finish the proof. □

The following is a consequence of Proposition 8.13 for functions on $\mathscr{P}_k$.

Proposition 8.14. *Let $k \in [1,\infty)$. Then for any $f \in C_K^{E,1}(\mathscr{P}_k)$ and $\mu, \nu \in \mathscr{P}_k$,*

$$\lim_{s \downarrow 0} \frac{f((1-s)\mu + s\nu) - f(\mu)}{s} = \int_{\mathbb{R}^d} \{\tilde{D}^E f(\mu)(x)\}(\nu - \mu)(\mathrm{d}x).$$

Proof. To apply Proposition 8.13, we extend a function f on $\mathscr{P}_k$ to $\tilde{f}$ on $\mathfrak{M}_k$ by letting

$$\tilde{f}(\mu) = h(\mu(\mathbb{R}^d))f(\mu/\mu(\mathbb{R}^d)), \quad \mu \in \mathfrak{M}_k,$$

where $h \in C_0^\infty(\mathbb{R})$ with support contained by $[\frac{1}{4}, 2]$ and $h(r) = 1$ for $r \in [\frac{1}{2}, \frac{3}{2}]$. It is easy to see that

$$f((1-s)\mu + s\nu) = \tilde{f}((1-s)\mu + s\nu), \quad s \in [0,1], \mu, \nu \in \mathscr{P}_k,$$

and $f \in C_K^{E,1}(\mathscr{P}_k)$ implies that $\tilde{f} \in C_K^{E,1}(\mathfrak{M}_k)$ and

$$D^E \tilde{f}(\mu) = \tilde{D}^E f(\mu), \quad \mu \in \mathscr{P}.$$

Then the desired formula is implied by Proposition 8.13 with $r = 0$. □

8.3 Links of Intrinsic and Extrinsic Derivatives

Theorem 8.15. *Let $k \in [1, \infty)$.*

(1) *Let $f \in C^{E,1,1}(\mathfrak{M}_k)$. Then f is intrinsically differentiable and*

$$Df(\mu)(x) = \nabla\{D^E f(\mu)(\cdot)\}(x), \quad (x, \mu) \in \mathbb{R}^d \times \mathfrak{M}_k. \quad (8.3.1)$$

When $k \in [1, 2]$ and $f \in C_B^{E,1,1}(\mathfrak{M}_k)$, we have $f \in C^1(\mathfrak{M}_k)$.

(2) *If $f \in C^1(\mathfrak{M}_k)$, then for any $s \geqslant 0$, $f(\mu + s\delta_\cdot) \in C^1(\mathbb{R}^d)$ with*

$$\nabla f(\mu + s\delta_\cdot)(x) = sDf(\mu + s\delta_x)(x), \quad x \in \mathbb{R}^d, s \geqslant 0. \quad (8.3.2)$$

Consequently,

$$Df(\mu)(x) = \lim_{s \downarrow 0} \frac{1}{s} \nabla f(\mu + s\delta_\cdot)(x), \quad (x, \mu) \in \mathbb{R}^d \times \mathfrak{M}_k. \quad (8.3.3)$$

Proof. In the following, we prove assertions (1) and (2), respectively.

(a) **Proof of assertion (1).** We first prove (8.3.1) for $f \in C^{E,1,1}(\mathfrak{M}_k)$. Let $v \in T_{\mu,k}$, and simply denote

$$\phi_{\varepsilon v} := \mathrm{id} + \varepsilon v, \quad \varepsilon \geqslant 0.$$

Since any $\mu \in \mathfrak{M}_k$ can be approximated by those having smooth and strictly positive density functions with respect to the volume measure

$\mathrm{d}x$, by the argument leading to (8.2.5), it suffices to show that for any $\mu \in \mathfrak{M}_k$ satisfying

$$\mu(\mathrm{d}x) = \rho(x)\,\mathrm{d}x \text{ for some } \rho \in C_b^\infty(\mathbb{R}^d),\ \inf \rho > 0, \tag{8.3.4}$$

there exists a constant $\varepsilon_0 > 0$ such that

$$\begin{aligned} &f(\mu \circ \phi_{\varepsilon v}^{-1}) - f(\mu) \\ &= \int_0^\varepsilon \mathrm{d}r \int_{\mathbb{R}^d} \langle \nabla\{D^E f(\mu \circ \phi_{rv}^{-1})\}, v\rangle \,\mathrm{d}(\mu \circ \phi_{rv}^{-1}), \quad \varepsilon \in (0, \varepsilon_0). \end{aligned} \tag{8.3.5}$$

Firstly, there exists a constant $\varepsilon_0 > 0$ such that

$$\rho_\varepsilon^v := \frac{\mathrm{d}(\mu \circ \phi_{\varepsilon v}^{-1})}{\mathrm{d}\mu}, \quad \dot\rho_\varepsilon^v := \lim_{s \downarrow 0} \frac{\rho_{\varepsilon+s}^v - \rho_\varepsilon^v}{s}$$

exist in $C_b(\mathbb{R}^d)$ and are uniformly bounded and continuous in $\varepsilon \in [0, \varepsilon_0]$. Next, by Proposition 8.12, we have

$$\begin{aligned} &f(\mu \circ \phi_{\varepsilon v}^{-1}) - f(\mu) \\ &= \int_0^\varepsilon \mathrm{d}r \int_{\mathbb{R}^d} \{D^E f(\mu \circ \phi_{rv}^{-1})\} \dot\rho_r^v \,\mathrm{d}\mu, \quad \varepsilon \in [0, \varepsilon_0]. \end{aligned} \tag{8.3.6}$$

To calculate $\dot\rho_r^v$, we note that for any $g \in C_0^\infty(\mathbb{R}^d)$,

$$\begin{aligned} \frac{\mathrm{d}}{\mathrm{d}r}\{g \circ \phi_{rv}\} &= \langle \nabla g(\phi_{rv}), v(\phi_{rv})\rangle \\ &= \langle \nabla g, v\rangle(\phi_{rv}), \quad r \geqslant 0, \end{aligned}$$

which is smooth and bounded in $(r, x) \in [0, \varepsilon_0] \times \mathbb{R}^d$. So,

$$\begin{aligned} &\int_{\mathbb{R}^d} g \dot\rho_r^v \,\mathrm{d}\mu = \int_{\mathbb{R}^d} g \lim_{s \downarrow 0} \frac{\rho_{r+s}^v - \rho_r^v}{s} \,\mathrm{d}\mu \\ &= \lim_{s \downarrow 0} \frac{1}{s} \int_{\mathbb{R}^d} g \,\mathrm{d}\{\mu \circ \phi_{(r+s)v}^{-1} - \mu \circ \phi_{rv}^{-1}\} \\ &= \lim_{s \downarrow 0} \frac{1}{s} \int_{\mathbb{R}^d} \{g \circ \phi_{(r+s)v} - g \circ \phi_{rv}\} \,\mathrm{d}\mu = \int_{\mathbb{R}^d} \frac{\mathrm{d}}{\mathrm{d}r}(g \circ \phi_{rv}) \,\mathrm{d}\mu \end{aligned}$$

$$
\begin{aligned}
&= \int_{\mathbb{R}^d} \langle \nabla g, v\rangle \circ \phi_{rv}\, \mathrm{d}\mu = \int_{\mathbb{R}^d} \langle \nabla g, v\rangle\, \mathrm{d}(\mu\circ\phi_{\varepsilon c}^{-1}) \\
&= -\int_{\mathbb{R}^d} \big\{ g\, \mathrm{div}_{\mu\circ\phi_{rv}^{-1}}(v)\big\}\, \mathrm{d}(\mu\circ\phi_{rv}^{-1}) \\
&= -\int_{\mathbb{R}^d} g\big\{ \mathrm{div}_{\mu\circ\phi_{rv}^{-1}}(v)\rho_r^v\big\}\, \mathrm{d}\mu, \quad g \in C_0^\infty(\mathbb{R}^d),
\end{aligned}
$$

where $\mathrm{div}_{\mu\circ\phi_{rv}^{-1}}(v) = \mathrm{div}(v) + \langle v, \nabla \log(\rho_r^v \rho)\rangle$. This implies

$$
\dot\rho_r^v = -\mathrm{div}_{\mu\circ\phi_{rv}^{-1}}(v)\rho_r,
$$

so that the integration by parts formula and

$$
\rho_r^v \mu = \mu\circ\phi_{rv}^{-1}
$$

lead to

$$
\begin{aligned}
&\int_{\mathbb{R}^d} \big\{ D^E f(\mu\circ\phi_{rv}^{-1})\big\} \dot\rho_r^v\, \mathrm{d}\mu \\
&\quad = -\int_{\mathbb{R}^d} \big\{ D^E f(\mu\circ\phi_{rv}^{-1})\big\} \mathrm{div}_{\mu\circ(\phi_{rv}^{-1}}(v)\; \mathrm{d}(\mu\circ\phi_{rv}^{-1}) \\
&\quad = \int_{\mathbb{R}^d} \big\langle \nabla\{ D^E f(\mu\circ\phi_{rv}^{-1})\}, v\big\rangle\, \mathrm{d}(\mu\circ\phi_{rv}^{-1}).
\end{aligned}
$$

Combining this with (8.3.6), we prove (8.3.5).

Now, let $k \in [1,2]$, we intend to verify the L-differentiability of f. For any $\mu \in \mathfrak{M}_k$ and $v \in T_{\mu,2}$ with $\mu(|v|^2) \leqslant 1$, we have

$$
\sup_{s\in[0,1]} (\mu\circ\phi_{sv}^{-1})(|\cdot|^k) = \mu(|\phi_{sv}|^k) \leqslant 2\mu(|\cdot|^k + |v|^k) < \infty.
$$

Then there exists a constant $K > 0$ such that

$$
\sup_{s\in[0,1],\mu(|v|^2)\leqslant 1} (\mu\circ(\phi_{sv})^{-1} + \mu)(|\cdot|^k) \leqslant K. \tag{8.3.7}
$$

So, by Proposition 8.13, we obtain

$$
\begin{aligned}
& f(\mu \circ (\mathrm{id} + v)^{-1}) - f(\mu) \\
&= \int_0^1 \left\{ \frac{\mathrm{d}}{\mathrm{d}r} f(r\mu \circ (\phi_{rv})^{-1} + (1-r)\mu) \right\} \mathrm{d}r \\
&= \int_0^1 \mathrm{d}r \int_{\mathbb{R}^d} (D^E f)(r\mu \circ \phi_{rv}^{-1} + (1-r)\mu)\, \mathrm{d}(\mu \circ \phi_v^{-1} - \mu) \\
&= \int_0^1 \mathrm{d}r \int_{\mathbb{R}^d} \Big\{ (D^E f)(r\mu \circ \phi_{rv}^{-1} + (1-r)\mu)(\phi_v(x)) \\
&\qquad\qquad - (D^E f)(r\mu \circ \phi_{rv}^{-1} + (1-r)\mu)(x) \Big\} \mu(\mathrm{d}x) \\
&= \int_0^1 \mathrm{d}r \int_{\mathbb{R}^d} \mu(\mathrm{d}x) \int_0^1 \langle \nabla \{ (D^E f)(r\mu \circ \phi_{rv}^{-1} \\
&\quad + (1-r)\mu) \}(\phi_{sv}(x)), v(x) \rangle \, \mathrm{d}s.
\end{aligned}
$$

Thus,

$$
\begin{aligned}
I_v &:= \frac{|f(\mu \circ \phi_v^{-1}) - f(\mu) - \int_{\mathbb{R}^d} \langle \nabla \{ D^E f(\mu) \}, v \rangle \, \mathrm{d}\mu|^2}{\mu(|v|^2)} \\
&\leqslant \int_{[0,1]^2 \times \mathbb{R}^d} |\nabla \{ (D^E f)(r\mu \circ \phi_{rv}^{-1} + (1-r)\mu) \}(\phi_{sv}(x)) \\
&\quad - \nabla \{ D^E f(\mu) \}(x)|^2 \, \mathrm{d}r \, \mathrm{d}s \mu(\mathrm{d}x).
\end{aligned}
$$

By (8.3.7), as $\|v\|_{L^2(\mu)} \to 0$, we have $\phi_{sv}(x) \to x$ μ-a.e. and $\mu \circ \phi_{sv}^{-1} \to \mu$ in $\mathfrak{M}_k$ for any $s \geqslant 0$. Combining these with (8.3.7), we may apply the dominated convergence theorem to derive $I_v \to 0$ as $\|v\|_{L^2(\mu)} \to 0$. Therefore, f is L-differentiable.

(b) **Proof of assertion (2).** It suffices to prove (8.3.2). Let $f \in C^1(\mathfrak{M}_k)$. We first prove the formula for $\mu \in \mathfrak{M}_k$ and $x \in \mathbb{R}^d$ with $\mu(\{x\}) = 0$ and then extend to the general situation.

Let $\mu(\{x\}) = 0$. In this case, for any $v_0 \in T_x M$, let $v = 1_{\{x\}} v_0$. Then

$$
\phi_{rv}(z) = \begin{cases} z & \text{if } z \neq x, \\ x + rv_0 & \text{if } z = x. \end{cases}
$$

By $\mu(\{x\}) = 0$, we have

$$(\mu + s\delta_x) \circ \phi_{rv}^{-1} = \mu + s\delta_{x+rv_0}. \tag{8.3.8}$$

Since v can be approximated in $L^2(\mu + s\delta_x)$ by smooth vector fields with compact support, the L-differentiability of f and $\mu(\{x\}) = 0$ imply

$$\begin{aligned}
&\lim_{r\downarrow 0} \frac{f((\mu + s\delta_x) \circ \phi_{rv}^{-1}) - f(\mu + s\delta_x)}{r} \\
&\quad = \int_{\mathbb{R}^d} \langle Df(\mu + s\delta_x), v\rangle \,\mathrm{d}(\mu + s\delta_x)) \\
&\quad = s\langle Df(\mu + s\delta_x)(x), v_0\rangle.
\end{aligned}$$

Combining this with (8.3.8), we obtain

$$\begin{aligned}
&\lim_{r\downarrow 0} \frac{f(\mu + s\delta_{x+rv_0}) - f(\mu + s\delta_x)}{r} \\
&\quad = s\langle Df(\mu + s\delta_x)(x), v_0\rangle.
\end{aligned}$$

This implies that $f(\mu + s\delta_\cdot)$ is differentiable at point x and (8.3.2) holds.

In general, for any $v_0 \in T_xM$, there exists $r_0 > 0$ such that v_0 extends to a smooth vector field v on $B(x, r_0)$ by parallel displacement, i.e. $v(x)$ is the parallel displacement along the minimal geodesic from x to z. Since $\mu(\{x + \theta v_0\}) = 0$ for a.e. $\theta \geqslant 0$, by the continuity of f and the formula (8.3.2) for $\mu(\{x\}) = 0$ proved above, we obtain

$$\begin{aligned}
&\frac{f(\mu + s\delta_{x+rv_0}) - f(\mu + s\delta_x)}{r} \\
&\quad = \frac{1}{r}\int_0^r \frac{\mathrm{d}}{\mathrm{d}\theta} f(\mu + s\delta_{x+\theta v_0}) \,\mathrm{d}\theta \\
&\quad = \frac{1}{r}\int_0^r \big\langle \nabla f(\mu + s\delta_\cdot)(x + \theta v_0), v(x + \theta v_0)\big\rangle \,\mathrm{d}\theta \\
&\quad = \frac{s}{r}\int_0^r \big\langle Df(\mu + s\delta_\cdot)(x + \theta v_0), v(x + \theta v_0)\big\rangle \,\mathrm{d}\theta, \quad r \in (0, r_0).
\end{aligned}$$

By the continuity of Df, with $r \downarrow 0$, this implies (8.3.2). □

Theorem 8.15 implies $C_B^{E,1,1}(\mathfrak{M}_k) \subset C^1(\mathfrak{M}_k)$ for $k \in [1,2]$. However, a function $f \in C^1(\mathfrak{M}_k)$ is not necessarily extrinsically differentiable. For instance, let $\psi \in C([0,\infty))$ but not differentiable, and let $f(\mu) = \psi(\mu(\mathbb{R}^d))$. Then $f(\mu + s\delta_x) = \psi(\mu(\mathbb{R}^d) + s)$ which is not differentiable in s, so that f is not extrinsically differentiable. But it is easy to see that $f \in C^1(\mathfrak{M}_k)$ with $Df(\mu) = 0$. Of course, this counter-example does not work for functions on $\mathscr{P}_k$.

By extending a function on $\mathscr{P}_k$ to $\mathfrak{M}_k$, we may apply Theorem 8.15 to establish the corresponding link for functions on $\mathscr{P}_k$. As an application, we will present derivative formula for the distributions of random variables.

For $s_0 > 0$ and a family of $\mathbb{R}^d$-valued random variables $\{\xi_s\}_{s\in[0,s_0)}$ on a probability space $(\Omega, \mathscr{F}, \mathbb{P})$, we say that $\dot{\xi}_0 := \frac{\mathrm{d}}{\mathrm{d}s}\xi_s\big|_{s=0}$ exists in $L^q(\mathbb{P})$ for some $q \geqslant 1$ if $\dot{\xi}_0 \in \mathscr{R}_k$ and

$$\lim_{s\downarrow 0} \mathbb{E}\left|\frac{\xi_s - \xi_0}{s} - \dot{\xi}_0\right|^q = 0. \tag{8.3.9}$$

Corollary 8.16. *Let $k \in [1,\infty)$.*

(1) *Let $f \in C^{E,1,1}(\mathscr{P}_k)$. Then f is intrinsically differentiable and*

$$Df(\mu)(x) = \nabla\{\tilde{D}^E f(\mu)(\cdot)\}(x), \quad (x,\mu) \in \mathbb{R}^d \times \mathscr{P}_k. \tag{8.3.10}$$

When $k \leqslant 2$ and $f \in C_B^{E,1,1}(\mathscr{P}_k)$, we have $f \in C^1(\mathscr{P}_k)$.

(2) *If $f \in C^{E,1}(\mathscr{P}_k)$, then $f((1-s)\mu + s\delta_\cdot) \in C^1(\mathbb{R}^d)$ with*

$$\nabla f((1-s)\mu + s\delta_\cdot)(x) = sDf((1-s)\mu + s\delta_x)(x), \quad x \in \mathbb{R}^d. \tag{8.3.11}$$

Consequently,

$$Df(\mu)(x) = \lim_{s\downarrow 0} \frac{1}{s}\nabla f((1-s)\mu + s\delta_\cdot)(x), \quad f \in C^{E,1}(\mathscr{P}_k),$$
$$(x,\mu) \in \mathbb{R}^d \times \mathfrak{M}. \tag{8.3.12}$$

(3) *Let $\{\xi_s\}_{s\in[0,s_0)}$ be random variables on M with $\mathscr{L}_{\xi_s} \in \mathscr{P}_k$ continuous in s, such that $\dot{\xi}_0 := \frac{\mathrm{d}}{\mathrm{d}s}\xi_s\big|_{s=0}$ exists in $L^q(\Omega \to TM; \mathbb{P})$ for some $q \geqslant 1$. Then*

$$\lim_{s\downarrow 0} \frac{f(\mathscr{L}_{\xi_s}) - f(\mathscr{L}_{\xi_0})}{s} = \mathbb{E}\langle Df(\mathscr{L}_{\xi_0})\}(\xi_0), \dot{\xi}_0\rangle \tag{8.3.13}$$

holds for any $f \in C^{E,1,1}(\mathscr{P}_k)$ *such that for any compact set* $\mathscr{K} \subset \mathscr{P}_k$,

$$\sup_{\mu\in\mathscr{K}} |\nabla\{\tilde{D}^E f(\mu)\}|(x) \leqslant C(1+|x|)^{\frac{p(q-1)}{q}}, \quad x \in \mathbb{R}^d \qquad (8.3.14)$$

holds for some constant $C > 0$.

Proof. To apply Theorem 8.15, we extend a function f on $\mathscr{P}_k$ to $\tilde{f}$ on $\mathfrak{M}_k$ as in the proof of Proposition 8.14, i.e. by letting

$$\tilde{f}(\mu) = h(\mu(\mathbb{R}^d)) f\left(\frac{\mu}{\mu(\mathbb{R}^d)}\right), \quad \mu \in \mathfrak{M}_k,$$

where $h \in C_0^\infty(\mathbb{R})$ with support contained in $[\frac{1}{4}, 2]$ and

$$h(r) = 1 \ \text{ for } r \in \left[\frac{1}{2}, \frac{3}{2}\right].$$

It is easy to see that

$$f((1-s)\mu + s\nu) = \tilde{f}((1-s)\mu + s\nu), \quad s \in [0,1], \mu, \nu \in \mathscr{P}_k,$$

and $f \in C^{E,1,1}(\mathscr{P}_k)$ implies that $\tilde{f} \in C^{E,1,1}(\mathfrak{M}_k)$ and

$$D^E \tilde{f}(\mu) = \tilde{D}^E f(\mu), \quad Df(\mu) = D\tilde{f}(\mu), \quad \mu \in \mathscr{P}.$$

Then Corollary 8.16(1)–(4) follow from the corresponding assertions in Theorem 8.15 with $\tilde{f}$ replacing f.

Finally, since $f \in C^{E,1,1}(\mathscr{P}_k)$ and

$$\nabla\{\tilde{D}^E f(\mu)\} = \nabla\{D^E \tilde{f}\}(\mu) = Df(\mu), \quad \mu \in \mathscr{P}_k,$$

(8.3.13) follows from Theorem 8.7. □

8.4 Gaussian Measures on $\mathscr{P}_2$ and $\mathfrak{M}$

The Gaussian measure, also called normal distribution, plays a key role in probability theory and related analysis. For instance, by the central limit theorem, the renormalization partial sum of i.i.d. random variables converges weakly to the standard Gaussian measure. In this section, we introduce Gaussian measures on $\mathscr{P}_2$ and $\mathfrak{M}$ as images of Gaussian distributions on Hilbert spaces.

8.4.1 Gaussian measure on Hilbert space

Let $\mathbb{H}$ be a separable Hilbert space, with orthonormal basis (ONB for short) $\{e_i\}_{i\geqslant 1}$. Let $(L, \mathscr{D}(L))$ be a positive definite self-adjoint operator on $\mathbb{H}$ with

$$Le_i = \alpha_i e_i, \quad i \geqslant 1$$

for positive constants $\{\alpha_i\}_{i\geqslant 1}$ satisfying $\alpha_i \downarrow 0$ as $i \uparrow \infty$ and

$$\sum_{i=1}^{\infty} \alpha_i < \infty.$$

Definition 8.17. Let $\{\xi_i\}_{i\geqslant 1}$ be i.i.d. random variables with standard normal distribution $N(0,1)$. Then

$$\xi = \sum_{i=1}^{\infty} \alpha_i^{\frac{1}{2}} \xi_i e_i$$

is called a Gaussian random variable on $\mathbb{H}$, whose distribution

$$G_L(A) := \mathbb{P}(\xi \in A), \quad A \in \mathscr{B}(\mathbb{H})$$

is called Gaussian measure on $\mathbb{H}$ with covariance operator L, denoted by $G_L = N(0, L)$.

In the following, we introduce the integration by parts formula for G_L. To this end, we introduce the class $C_b^1(\mathbb{H})$ of functions on $\mathbb{H}$.

Definition 8.18. A function f on $\mathbb{H}$ is called Gâdeaux differentiable if for any $x \in \mathbb{H}$,

$$\mathbb{H} \ni v \mapsto \nabla_v f(x) := \lim_{\varepsilon \downarrow 0} \frac{f(x+\varepsilon v) - f(x)}{\varepsilon} \in \mathbb{R}$$

is a well-defined linear functional. In this case, the unique element $\nabla f(x) \in \mathbb{H}$ such that

$$\langle \nabla f(x), v \rangle_{\mathbb{H}} = \nabla_v f(x), \quad v \in \mathbb{H}$$

is called the Gâdeaux derivative of f at point x.

If f has Gâdaeux derivative and

$$\lim_{\|v\|_{\mathbb{H}} \downarrow 0} \frac{f(x+v) - f(x) - \nabla_v f(x)}{\|v\|_{\mathbb{H}}} = 0, \quad x \in \mathbb{H},$$

then f is called Fréchet differentiable.

If f is Fréchet differentiable and $\nabla f : \mathbb{H} \to \mathbb{H}$ is continuous, we denote $f \in C^1(\mathbb{H})$. If, moreover, $\|\nabla f\|_{\mathbb{H}} + |f|$ is bounded on $\mathbb{H}$, we denote $f \in C_b^1(\mathbb{H})$.

Next, we introduce the divergence of vector fields on $\mathscr{H}$.

Definition 8.19. A vector field is a measurable map

$$v : \mathscr{H} \to \mathscr{H}.$$

We denote $v \in C_b^1(\mathscr{H}; \mathscr{H})$ for each $i \geqslant 1$, we have

$$v_i := \langle v, e_i \rangle_{\mathscr{H}} \in C_b^1(\mathscr{H}),$$

and there exists a constant $c > 0$, such that

$$|v| + \sum_{i=1}^{\infty} |\nabla_{e_i} v_i| \leqslant c.$$

In this case,

$$\nabla^* v := \sum_{i=1}^{\infty} \nabla_{e_i} v_i \in C(\mathscr{H})$$

is called the divergence of v.

Theorem 8.20. *Let $f \in C_b^1(\mathbb{H})$ and $v \in C_b^1(\mathbb{H}; \mathbb{H})$ such that $\sum_{i=1}^{\infty} |v_i| \alpha_i^{-\frac{1}{2}}$ is bounded. Then*

$$\int_{\mathbb{H}} \langle v, \nabla f \rangle_{\mathbb{H}} \, \mathrm{d}G_L = \int_{\mathscr{H}} \left(\sum_{i=1}^{\infty} \frac{r_i v_i}{\alpha_i} - \nabla^* v \right) f \, \mathrm{d}\, G_L.$$

Proof. We first formulate G_L by using the coordinates

$$\Phi : \mathbb{H} \ni x \mapsto (\langle x, e_i \rangle_{\mathbb{H}})_{i \geqslant 1} \in \ell^2,$$

where

$$\ell^2 := \left\{ r = (r_i)_{i \geqslant 1} \in \mathbb{R}^{\mathbb{N}} : \sum_{i=1}^{\infty} r_i^2 < \infty \right\}$$

is a Hilbert space. Let

$$\Lambda_i(\mathrm{d}r_i) = \frac{1}{\sqrt{2\alpha_i \pi}} \mathrm{e}^{-\frac{r_i^2}{2\alpha_i}} \, \mathrm{d}r_i, \quad i \geqslant 1$$

be the centered normal distribution with variance α_i. Then the product measure

$$\Lambda := \prod_{i=1}^{\infty} \Lambda_i$$

is supported on ℓ^2, since

$$\int_{\mathbb{R}^{\mathbb{N}}} \left(\sum_{i=1}^{\infty} r_i^2 \right) \Lambda(\mathrm{d}r) = \sum_{i=1}^{\infty} \alpha_i^2 < \infty.$$

It is easy to see that

$$G_L = \Lambda \circ \Phi^{-1}.$$

By combining this with the integral transformation theorem (Theorem 3.27) and the integration by parts formula

$$\int_{\mathbb{R}} h(r_i) g'(r_i) \, \mathrm{d}\Lambda_i = \int_{\mathbb{R}} \left(\frac{r_i h(r_i)}{\alpha_i} - h'(r_i) \right) \Lambda_i(\mathrm{d}r_i), \quad h, g \in C_b^1(\mathbb{R}),$$

we finish the proof. □

8.4.2 Gaussian measures on $\mathscr{P}_2$

Let $\mu_0 \in \mathscr{P}_2$ such that the tangent space $\mathscr{H} := T_{\mu_0,2}$ is a separable Hilbert space. Let G_L be the Gaussian measure on $\mathscr{H}$. This measure induces a Gaussian measure on $\mathscr{P}_2$ under the map

$$\Phi_1 : \mathscr{H} \ni \phi \mapsto \mu_0 \circ \phi^{-1} \in \mathscr{P}_2.$$

Definition 8.21. We call

$$N_{\mu_0,L} := G_L \circ \Phi_1^{-1}$$

the Gaussian measure on $\mathscr{P}_2$ with parameter (μ_0, L).

By the chain rule Theorem 8.7, we have the following result.

Theorem 8.22. *Let $u, v \in C_b^1(\mathscr{P}_2)$. Then*

$$f := u \circ \Phi_1, \quad g := v \circ \Phi_1 \in C_b^1(T_{\mu_0,2})$$

and

$$\int_{\mathscr{P}_2} \langle Du(\mu), Dv(\mu) \rangle_{T_{\mu,2}} N_{\mu_0,L}(\mathrm{d}\mu) = \int_{T_{\mu_0,2}} \langle \nabla f, \nabla g \rangle_{T_{\mu_0,2}} \, \mathrm{d}G_L.$$

8.4.3 Gaussian measures on $\mathfrak{M}$

In this section, we construct Gaussian measures on $\mathfrak{M}$ supported on the subspace of absolutely continuous measures

$$\mathfrak{M}_{\mathrm{ac}} := \left\{ \mu \in \mathfrak{M} : \ \rho_\mu := \frac{\mathrm{d}\mu}{\mathrm{d}x} \text{ exists} \right\}.$$

In this case, we choose

$$\mathbb{H} = L^2(\mathbb{R}^d, \mathrm{d}x) := \left\{ h : \mathbb{R}^d \to \mathbb{R} \text{ is measurable}, \ \int_{\mathbb{R}^d} h(x)^2 \, \mathrm{d}x < \infty \right\}.$$

We then consider the image of the Gaussian measure G_L on $\mathbb{H}$ under the map

$$\Phi_2 : \mathbb{H} = L^2(\mathrm{d}x) \ni h \mapsto h(x)^2 \, \mathrm{d}x \in \mathfrak{M}_{\mathrm{ac}}.$$

Definition 8.23. We call

$$N_L := G_L \circ \Phi_2^{-1}$$

the Gaussian measure on $\mathfrak{M}$ (or $\mathfrak{M}_{\mathrm{ac}}$) with parameter L.

By Proposition 8.12 for $k = 0$ (see Exercise 5), we may prove the following result.

Theorem 8.24. *Let $u, v \in C_b^{E,1,1}(\mathfrak{M})$, and let $\mathbb{H} = L^2(\mathbb{R}^d, \mathrm{d}x)$. Then*

$$f := u \circ \Phi_2, \quad g := v \circ \Phi_2 \in C_b^1(\mathbb{H})$$

and

$$\int_{\mathscr{P}_2} \langle D^E u(\mu), D^E v(\mu) \rangle_{L^2(\mu)} \, N_L(\mathrm{d}\mu) = \int_{\mathbb{H}} \langle \nabla f, \nabla g \rangle_{\mathbb{H}} \, \mathrm{d}G_L.$$

8.5 Exercises

1. Prove that $\mathfrak{M}_k$ is a Polish space.

2. Let

$$f(\mu) = g(\mu(h_1), \ldots, \mu(h_n))$$

for some $n \geqslant 1, g \in C^1(\mathbb{R}^n)$ and $h_i \in C_b^1(\mathbb{R}^d), 1 \leqslant i \leqslant n$. Prove that $f \in C_b^1(\mathscr{P}_k)$ and

$$Df(\mu) = \sum_{i=1}^{d} (\partial_i g)(\mu(h_1), \ldots, \mu(h_n)) \nabla h_i.$$

This type of functions are called C_b^1-cylindrical functions on $\mathscr{P}_k$.

3. Let $\mu_0 \in \mathscr{P}_k$ be absolutely continuous with respect to the Lebesgue measure. By [2, Theorem 6.2.10], for any $\mu \in \mathscr{P}_k$, there exists a unique $\phi_\mu \in T_{\mu_0,k}$ such that

$$\mu = \mu_0 \circ (\mathrm{id} + \phi_\mu)^{-1}, \quad \mathbb{W}_k(\mu_0, \mu) = (\mu_0(|\phi_\mu|^k))^{\frac{1}{k}}.$$

Please use this assertion and the chain rule to prove the following: If $f \in C^1(\mathscr{P}_k)$ satisfies (8.1.12) for all $\mu \in \mathscr{P}_k$ with some constant $c(\mu) > 0$, then for any $\mu \in \mathscr{P}_k$,

$$f(\mu) - f(\mu_0) = \int_0^1 \langle (Df)(\mu_0 \circ (\mathrm{id} + r\phi_\mu)^{-1}), \phi_\mu \rangle_{L^2(\mu_0)} \, \mathrm{d}r.$$

4. Let $k \in [1, \infty)$. For two probability measures $\mu, \nu \in \mathscr{P}_k$, define the k-variational distance

$$\|\mu - \nu\|_{k,\mathrm{var}} := \sup_{|f| \leqslant 1 + |\cdot|^k} |\mu(f) - \nu(f)|.$$

For any $f \in C_K^{E,1,1}(\mathscr{P}_k)$ such that

$$|\tilde{D} f(\mu)(x)| \leqslant 1 + |x|^k, \quad \mu \in \mathscr{P}_k, \ x \in \mathbb{R}^d,$$

prove

$$|f(\mu) - f(\nu)| \leqslant \|\mu - \nu\|_{k,\mathrm{var}}.$$

5. For $k \in [0, 1]$, let $\mathscr{P}_k$, $\mathfrak{M}_k$, D^E, $\tilde{D}^E$, $C^{E,1,1}(\mathfrak{M}_k)$, $C_K^{E,1,1}(\mathscr{P}_k)$ and $C_K^{E,1,1}(\mathfrak{M}_k)$ be defined as before. Prove that Propositions 8.12–8.14 still hold. Can we also define the directional intrinsic derivative and intrinsic derivative on $\mathscr{P}_k$ and $\mathfrak{M}_k$ for $k \in [0, 1]$?

6. Let $k \in [0, \infty)$. Prove that there exists a constant $c > 0$ such that

$$\|\mu - \nu\|_{\mathrm{var}} + \mathbb{W}_k(\mu, \nu)^{1 \vee k} \leqslant c \|\mu - \nu\|_{k,\mathrm{var}}, \quad \mu, \nu \in \mathscr{P}_k.$$

Moreover, when $k > 1$, find counter example such that for any constant $c > 0$, the inequality

$$\mathbb{W}_k(\mu, \nu) \leqslant c\|\mu - \nu\|_{k,\mathrm{var}}, \quad \mu, \nu \in \mathscr{P}_k$$

does not hold.

7. Prove Theorem 8.22.

8. Prove Theorem 8.24.

Bibliography

[1] Billingsley, P. *Probability and Measure.* 3rd Ed. John Wiley and Sons, New York, 1995.

[2] Bogachev, I. *Measure Theory* (Vol. I). Springer, Berlin, 2007.

[3] Chen, Mu-Fa. *From Markov Chains to Non-equilibrium Particle Systems.* 2nd Ed. World Scientific, River Edge, NJ, 2004.

[4] Dudley, R. *Real Analysis and Probability.* Wadsworth, Pacific Grove, CA, 1989.

[5] Durrett, R. *Probability: Theory and Examples.* 5th Ed. Cambridge University Press, Cambridge, 2019.

[6] Kallenberg, O. *Foundations of Modern Probability.* 3rd Ed. Springer, Cham, 2021.

[7] Feller, W. *An Introduction to Probability and Its Applications* (Vol. I, II). John Wiley and Sons, London, 1971.

[8] Loéve, M. *Probability Theory* II. 4th Ed. Springer-Verlag, New York-Heidelberg, 1978.

[9] Neveu, J. *Mathematical Foundations of the Calculus of Probability.* Holden-Day, San Francisco, Calif.-London-Amsterdam, 1965.

[10] Rachev, S. *The Monge–Kantorovich Mass Transference Problem and Its Stochastic Applications.* Theory of Probability and Applications, Vol. XXIX, 1985, 647–676.

[11] Reed, M., Simon, B. *Method of Modern Mathematical Physics* (Vol. I). Academic Press, New York, 1980.

[12] Shiryayev, A. *Probability.* Springer-Verlag, New York, 1984.

[13] Villani, C. *Optimal Transport.* Springer-Verlag, Berlin, 2009.

[14] Wang, Jaigang. *Foundation of Modern Probability Theory* (in Chinese). 2nd Ed. Fudan University Press, Shanghai, 2005.

[15] Yan, Jiaan. *Lectures on Measure Theory* (in Chinese). 2nd Ed. Science Press, Beijing, 2004.

[16] Yan, Shi-Jian, Liu, Xiu-Fang. *Measure and Probability* (in Chinese). Beijing Norm University Press, Beijing, 2003.
[17] Yan, Shi-Jian, Wang, Jun-Xiang, Liu, Xiu-Fang. *Foundation of Probability Theory* (in Chinese). 2nd Ed. Science Press, Beijing, 2009.
[18] Yosida, K. *Functional Analysis.* 6th Ed. Springer-Verlag, Berlin, 1980.
[19] Zhang, Gongqing, Lin, Yuanqu. *Lectures on Functional Analysis* (Vol I) (in Chinese). Peking University Press, Beijing, 1990.
[20] Zhang, Gongqing, Guo, Maozheng. *Lectures on Functional Analysis* (Vol II) (in Chinese). Peking University Press, Beijing, 2001.

Index

Printed in the USA
CPSIA information can be obtained
at www.ICGtesting.com
LVHW011716161124
796134LV00013B/21

* 9 7 8 9 8 1 1 2 9 8 8 5 1 *